S0-BRA-606

Reviews of Environmental Contamination and Toxicology

VOLUME 141

Reviews of Environmental Contamination and Toxicology

Continuation of Residue Reviews

Editor
George W. Ware

Editorial Board
F. Bro-Rasmussen, Lyngby, Denmark
D.G. Crosby, Davis, California, USA · H. Frehse, Leverkusen-Bayerwerk, Germany
H.F. Linskens, Nijmegen, The Netherlands · O. Hutzinger, Bayreuth, Germany
N.N. Melnikov, Moscow, Russia · M.L. Leng, Midland, Michigan, USA
D.P. Morgan, Oakdale, Iowa, USA · P. De Pietri-Tonelli, Milano, Italy
Annette E. Pipe, Burnaby, British Columbia, Canada
Raymond S.H. Yang, Fort Collins, Colorado, USA

Founding Editor
Francis A. Gunther

VOLUME 141

Springer-Verlag
New York Berlin Heidelberg London Paris
Tokyo Hong Kong Barcelona Budapest

Coordinating Board of Editors

GEORGE W. WARE, *Editor*
Reviews of Environmental Contamination and Toxicology

Department of Entomology
University of Arizona
Tucson, Arizona 85721, USA
(602) 299-3735 (phone and FAX)

HERBERT N. NIGG, *Editor*
Bulletin of Environmental Contamination and Toxicology

University of Florida
700 Experimental Station Road
Lake Alfred, Florida 33850, USA
(813) 956-1151; FAX (813) 956-4631

ARTHUR BEVENUE, *Editor*
Archives of Environmental Contamination and Toxicology

4213 Gann Store Road
Hixson, Tennessee 37343, USA
(615) 877-5418

Springer-Verlag
New York: 175 Fifth Avenue, New York, NY 10010, USA
Heidelberg: 69042 Heidelberg, Postfach 10 52 80, Germany

Library of Congress Catalog Card Number 62-18595.
Printed in the United States of America.

ISSN 0179-5953

© 1995 by Springer-Verlag New York, Inc.
All rights reserved. This work may not be translated or copied in whole or in part without the written permission of the publisher (Springer-Verlag, 175 Fifth Avenue, New York, NY 10010, USA), except for brief excerpts in connection with reviews or scholarly analysis. Use in connection with any form of information storage and retrieval, electronic adaptation, computer software, or by similar or dissimilar methodology now known or hereafter developed is forbidden.
The use of general descriptive names, trade names, trademarks, etc., in this publication, even if the former are not especially identified, is not to be taken as a sign that such names, as understood by the Trade Marks and Merchandise Marks Act, may accordingly be used freely by anyone.

ISBN 0-387-94453-2 Springer-Verlag New York Berlin Heidelberg

Foreword

International concern in scientific, industrial, and governmental communities over traces of xenobiotics in foods and in both abiotic and biotic environments has justified the present triumvirate of specialized publications in this field: comprehensive reviews, rapidly published research papers and progress reports, and archival documentations. These three international publications are integrated and scheduled to provide the coherency essential for nonduplicative and current progress in a field as dynamic and complex as environmental contamination and toxicology. This series is reserved exclusively for the diversified literature on "toxic" chemicals in our food, our feeds, our homes, recreational and working surroundings, our domestic animals, our wildlife and ourselves. Tremendous efforts worldwide have been mobilized to evaluate the nature, presence, magnitude, fate, and toxicology of the chemicals loosed upon the earth. Among the sequelae of this broad new emphasis is an undeniable need for an articulated set of authoritative publications, where one can find the latest important world literature produced by these emerging areas of science together with documentation of pertinent ancillary legislation.

Research directors and legislative or administrative advisers do not have the time to scan the escalating number of technical publications that may contain articles important to current responsibility. Rather, these individuals need the background provided by detailed reviews and the assurance that the latest information is made available to them, all with minimal literature searching. Similarly, the scientist assigned or attracted to a new problem is required to glean all literature pertinent to the task, to publish new developments or important new experimental details quickly, to inform others of findings that might alter their own efforts, and eventually to publish all his/her supporting data and conclusions for archival purposes.

In the fields of environmental contamination and toxicology, the sum of these concerns and responsibilities is decisively addressed by the uniform, encompassing, and timely publication format of the Springer-Verlag (Heidelberg and New York) triumvirate:

Reviews of Environmental Contamination and Toxicology [Vol. 1 through 97 (1962–1986) as Residue Reviews] for detailed review articles concerned with any aspects of chemical contaminants, including pesticides, in the total environment with toxicological considerations and consequences.

Bulletin of Environmental Contamination and Toxicology (Vol. 1 in 1966)
for rapid publication of short reports of significant advances and discoveries in the fields of air, soil, water, and food contamination and pollution as well as methodology and other disciplines concerned with the introduction, presence, and effects of toxicants in the total environment.

Archives of Environmental Contamination and Toxicology (Vol. 1 in 1973)
for important complete articles emphasizing and describing original experimental or theoretical research work pertaining to the scientific aspects of chemical contaminants in the environment.

Manuscripts for *Reviews* and the *Archives* are in identical formats and are peer reviewed by scientists in the field for adequacy and value; manuscripts for the *Bulletin* are also reviewed, but are published by photo-offset from camera-ready copy to provide the latest results with minimum delay. The individual editors of these three publications comprise the joint Coordinating Board of Editors with referral within the Board of manuscripts submitted to one publication but deemed by major emphasis or length more suitable for one of the others.

Coordinating Board of Editors

Preface

Worldwide, anyone keeping abreast of current events is exposed daily to multiple reports of environmental insults: global warming (greenhouse effect) in relation to atmospheric CO_2, nuclear and toxic waste disposal, massive marine oil spills, acid rain resulting from atmospheric SO_2 and NO_x, contamination of the marine *commons*, deforestation, radioactive contamination of urban areas by nuclear power generators, and the effect of free chlorine and chlorofluorocarbons in reduction of the earth's ozone layer. These are only the most prevalent topics. In more localized settings we are reminded of exposure to electric and magnetic fields; indoor air quality; leaking underground fuel tanks; increasing air pollution in our major cities; radon seeping from the soil into homes; movement of nitrates, nitrites, pesticides, and industrial solvents into groundwater; and contamination of our food and feed with bacterial toxins. Some of the newer additions to the vocabulary include *xenobiotic transport*, *solute transport*, *Tiers 1 and 2*, *USEPA to cabinet status*, and *zero-discharge*.

It then comes as no surprise that ours is the first generation of mankind to have become afflicted with the pervasive and acute fear of chemicals, appropriately named *chemophobia*.

There is abundant evidence, however, that virtually all organic chemicals are degraded or dissipated in our not-so-fragile environment, despite efforts by environmental ethicists and the media to persuade us otherwise. But for most scientists involved in reduction of environmental contaminants, there is indeed room for improvement in all spheres.

Environmentalism has become a global political force, resulting in multinational consortia emerging to control pollution and in the maturation of the environmental ethic. Will the new politics of the next century be a consortium of technologists and environmentalists or a progressive confrontation? These matters are of genuine concern to governmental agencies and legislative bodies around the world, for many chemical incidents have resulted from accidents and improper use.

For those who make the decisions about how our planet is managed, there is an ongoing need for continual surveillance and intelligent controls, to avoid endangering the environment, wildlife, and the public health. Ensuring safety-in-use of the many chemicals involved in our highly industrialized culture is a dynamic challenge, for the old established materials are continually being displaced by newly developed molecules more acceptable to environmentalists, federal and state regulatory agencies, and public health officials.

Adequate safety-in-use evaluations of all chemicals persistent in our air, foodstuffs, and drinking water are not simple matters, and they incorporate the judgments of many individuals highly trained in a variety of complex biological, chemical, food technological, medical, pharmacological, and toxicological disciplines.

Reviews of Environmental Contamination and Toxicology continues to serve as an integrating factor both in focusing attention on those matters requiring further study and in collating for variously trained readers current knowledge in specific important areas involved with chemical contaminants in the total environment. Previous volumes of *Reviews* illustrate these objectives.

Because manuscripts are published in the order in which they are received in final form, it may seem that some important aspects of analytical chemistry, bioaccumulation, biochemistry, human and animal medicine, legislation, pharmacology, physiology, regulation, and toxicology have been neglected at times. However, these apparent omissions are recognized, and pertinent manuscripts are in preparation. The field is so very large and the interests in it are so varied that the Editor and the Editorial Board earnestly solicit authors and suggestions of underrepresented topics to make this international book series yet more useful and worthwhile.

Reviews of Environmental Contamination and Toxicology attempts to provide concise, critical reviews of timely advances, philosophy, and significant areas of accomplished or needed endeavor in the total field of xenobiotics in any segment of the environment, as well as toxicological implications. These reviews can be either general or specific, but properly they may lie in the domains of analytical chemistry and its methodology, biochemistry, human and animal medicine, legislation, pharmacology, physiology, regulation, and toxicology. Certain affairs in food technology concerned specifically with pesticide and other food-additive problems are also appropriate subjects.

Justification for the preparation of any review for this book series is that it deals with some aspect of the many real problems arising from the presence of any foreign chemical in our surroundings. Thus, manuscripts may encompass case studies from any country. Added plant or animal pest-control chemicals or their metabolites that may persist into food and animal feeds are within this scope. Food additives (substances deliberately added to foods for flavor, odor, appearance, and preservation, as well as those inadvertently added during manufacture, packing, distribution, and storage) are also considered suitable review material. Additionally, chemical contamination in any manner of air, water, soil, or plant or animal life is within these objectives and their purview.

Normally, manuscripts are contributed by invitation, but suggested topics are welcome. Preliminary communication with the Editor is recommended before volunteered review manuscripts are submitted.

Department of Entomology   G.W.W.
University of Arizona
Tucson, Arizona

Table of Contents

Polybrominated Biphenyl and Diphenylether Flame Retardants: Analysis, Toxicity, and Environmental Occurrence

A.M.C.M. Pijnenburg,* J.W. Everts,*,** J. de Boer,†
and J.P. Boon‡

Contents

*National Institute for Coastal and Marine Management (RIKZ), Ministry of Transport, Public Works and Water Management, The Hague, the Netherlands.

**present address: FAO, BP 3300, Dakar, Sénégal.

†Netherlands Institute for Fisheries Research (RIVO), IJmuiden, the Netherlands.

‡Netherlands Institute for Sea Research (NIOZ), Texel, the Netherlands. This is publication no. 41 of the Applied Science project of BeWON.

© 1995 by Springer-Verlag New York, Inc.
Reviews of Environmental Contamination and Toxicology, Vol. 141.

I. Introduction

Polybromobiphenyls (PBBs) and polybromodiphenylethers (PBDEs) are presently being used as flame retardants in electronic equipment, plastics, building materials, and carpets. There are many standards and regulations applicable to flame retardants; those issued by the American Society for Testing of Materials (ASTM) alone account for more than one hundred (Arias 1992). The advantage of these compounds for industry is their high resistance toward acids, bases, heat, light, and reducing and oxidizing compounds. However, this high resistance becomes a great disadvantage when these compounds are discharged into the environment, where they persist for a long time. Furthermore, brominated dibenzofurans and dibenzodioxins may be formed when these flame retardants are heated (Watanabe and Tatsukawa 1990).

The basic formulas of the PBBs and PBDEs are shown in Figure 1. Throughout this chapter, the systematic numbering system developed by Ballschmiter and Zell (1980) for the 209 theoretically possible polychlorinated biphenyl (PCB) congeners has also been adopted for the corresponding PBB and PBDE congeners.

PCBs may now be encountered globally, and they have caused one of the major environmental problems as a result of their chronic and diffuse input into the environment (Tanabe 1988). The similarity in molecular structure of the PBBs and, to a lesser degree, the PBDEs, with the PCBs, give rise to great environmental concern. Much of the quantities of PBB and PBDEs produced will eventually reach the marine environment, where they are likely to accumulate because of their persistence or resistance to degradative processes. Because of their environmental properties, the continued release of PBBs and PBDEs represents an increasing risk to the environment. The purpose of this chapter is to present the existing data in a form that will lead to increased knowledge of this problem, eventually enhancing the proper measures.

II. Analysis

Because of the similarity of the character of PBBs and PBDEs to PCBs, the methods for analysis of PCBs, which are abundantly available in the literature, will also be applicable to analysis of PBBs and PBDEs. The necessary modifications are described herein.

A. Analysis of PBBs

Extraction and cleanup techniques for the analysis of PBBs in fatty tissues and sediments are identical to extraction and cleanup techniques for PCB analysis. Fehringer (1975) describes the extraction of PBBs with dichloromethane from dry animal food. Cleanup is performed with Florisil columns. Soxhlet extraction with dichloromethane/n-pentane, followed by

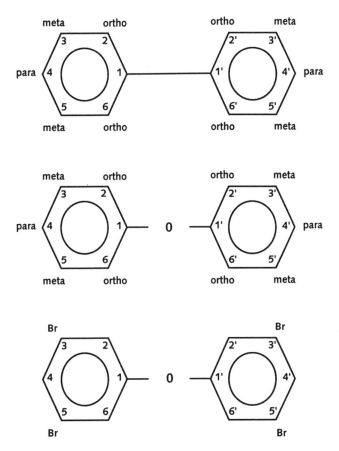

Fig. 1. Basic formulas of brominated fire retardants: (a) Polybromobiphenyls (PBBs); (b) Polybromodiphenylethers (PBDEs); (c) 3,3′,5,5′-tetrabromodiphenyl-ether.

cleanup through alumina columns and fractionation over SiO_2 columns, results in recoveries greater than 95%. Saponification (de Boer 1988a) may be an alternative, but decomposition of some PBB congeners may occur as in the case of PCBs (van der Valk and Dao 1988).

Gas chromatography (GC) is the obvious method for the final analysis of PBBs. Several methods using packed columns are described by Fehringer (1975), de Kok et al. (1977), Sweetman and Boettner (1982), Domino et al. (1980), Mulligan and Caruso (1980), and Erickson et al. (1980). The most frequently used stationary phases are SE-30, OV-1, and OV-17. Oven temperatures vary between 240° and 300°C. The PBBs elute after the PCBs, but higher chlorinated PCBs may interfere with lower brominated PBBs. Polychlorinated terphenyls (PCTs) may also interfere with PBBs

(Wester and de Boer 1993). The separation of PBBs on a capillary 10 m ×
0.25 mm i.d. OV-101 column is described by Farrell (1980).

For detection of peaks after separation, either an electron capture detector (GC-ECD) (Fehringer 1975; Domino et al. 1980; Farrell 1980; Sweetman and Boettner 1982) or a mass spectrometer (GC-MS) (de Kok et al.
1977; Erickson et al. 1980) is most frequently used. Mulligan and Caruso
(1980) used a plasma-emission detector. Because of the possible interfer-
ence of PCB and PCT congeners, GC-MS analysis will probably be the
most advantageous technique. The use of negative chemical ionization
(NCI) as ionization technique for the GC-MS analysis is advantageous
because of its high sensitivity for compounds with four or more bromine
atoms (Wester and de Boer 1993). The sensitivity of NCI for these compounds is approximately 10 times better than that of ECD. The use of
narrow bore (0.15 mm i.d.) capillary GC columns is advised to obtain the
required resolution.

As for all mixtures of chemical compounds with a similar structure, only
the quantification of individual brominated biphenyls (BB) congeners (instead of quantification on the basis of technical mixture equivalents) will allow a comparative study of the environmental occurrence, behavior, and toxicokinetics of PBBs. However, no such results have yet
been reported, although some individual BB congeners are available as
standards.

B. Analysis of PBDEs

Extraction and cleanup techniques for the analysis of PBDEs in fatty tissues
and sediments are similar to extraction and cleanup techniques for PCB
and PBB analysis (see Section IIA). Watanabe et al. (1987) describe PBDE
analysis in fish and sediments using an acetone/hexane extraction, fat removal by concentrated sulfuric acid, and cleanup on a Florisil column.
Andersson and Blomkist (1981) describe PBDE analysis in fish by the
subsequent use of a hexane/acetone and a hexane/diethylether extraction,
fat removal by concentrated sulfuric acid, and cleanup on a silica column.
The analysis of PBDEs described by de Boer (1989) is performed by soxhlet
extraction with pentane/dichloromethane, fat removal over alumina columns, and fractionation over silica. In contrast to the PCBs and the PBBs,
the more polar PBDEs elute in the second silica fraction together with some
organochlorine pesticides.

A multiresidue method including the analysis of PBDEs is described by
Jansson et al. (1991). The PBDEs are extracted from biological samples
with hexane/acetone and hexane/diethylether, treated with sulfuric acid
and cleaned by SX-3 Biobeads gel permeation and silica gel column chromatography. However, the mean recovery of 2,2′,4,4′-tetra-BDE for this
method is only 49%.

Both GC-ECD and GC-MS with electron impact (EI) or NCI may be used for the final analysis of PBDEs. On nonpolar or moderately polar capillary columns, 2,2',4,4'-tetra-BDE and 2,2',4,4',5-penta-BDE elute late in the gas chromatogram of the second silica fraction between p,p'-DDT and octachloronaphthalene. The plasticizer di(2-ethylhexyl)phthalate, originating from septa, for example, may coelute with 2,2',4,4'-tetra-BDE.

The technical mixture Bromkal 70 5DE is used as the external standard for quantification in most studies. For reasons described in section II A, there is a need for individual bromodiphenylethers as analytical standards, but they are not commercially available. Wolf and Rimkus (1985) described the synthesis of 2,2',4,4'-tetra-BDE for the analysis of this congener in fish.

III. Environmental Fate and Occurrence
A. Use

In the Netherlands, the annual consumption of PBBs and PBDEs is 250 and 2500 tonnes, respectively; in Sweden, it is 1400–2000 t/yr (Svensson and Hellsten 1989); and in Japan, between 1975 and 1987, the consumption increased from 2500 to 22,100 t/yr (Watanabe and Tatsukawa 1990).

B. Environmental Fate

Higher brominated compounds have a lower solubility in water than the corresponding chlorinated compounds. The volatility of PBBs is lower than the volatility of the corresponding PCBs. The vapor pressure (P_{vp}) of hexabromobiphenyl and hexachlorobiphenyl is 4.52×10^{-10} Pa and 3.94×10^{-8} Pa, respectively. The P_{vp} of decabromobiphenyl is $< 7.4 \times 10^{-4}$; and for decachlorobiphenyl, it is 5.31×10^{-10} Pa. Higher BBs, as in flame retardants, attach strongly to sediment close to discharge points. Lower BBs have a greater solubility in water and are more easily distributed in the aquatic environment (Watanabe and Tatsukawa 1990).

Both PBBs and PBDEs are slowly degraded in the environment. Since 1970, limited *in situ* reductive debromination of Firemaster mixture seemed to occur in Pine River sediments (St Louis, Michigan, U.S.). In 1988, sediment cores contained 10–12% non-Firemaster PBB compounds. It appeared that bromines were selectively removed from the *meta* and *para* positions.

Microorganisms isolated from Pine River sediment were capable of debrominating Firemaster PBB compounds. Organic cocontaminants, petroleum products, and heavy metals inhibited *in situ* debromination in the most contaminated Pine River sediments (Morris et al. 1993). According to Watanabe et al. (1986), decabromodiphenylether dissolved in hexane can be degraded to nona-, octa-, hepta-, and hexabromodiphenylethers. Ruzo et al. (1976) studied photodegradation of PBBs dissolved in hexane. PBBs with bromides at the *ortho* positions degrade most rapidly. Mills et al. (1985) carried out a photolysis experiment with hexa-BB components from PBB mixtures dissolved in hexane. It was found that these components could be transformed into tetra- and penta-BB components under the influence of UV light. Buser (1986) has described the formation of bromodibenzodioxins and bromodibenzofurans from PBDE by thermolysis.

Bioaccumulation and metabolism of PBBs and PBDEs in higher organisms are described in section IV.

C. Environmental Occurrence

Residues of PBDE found in samples from the environment are listed in Table 1. In sediments of the Baltic Sea, an increasing trend in the concentration of PBDEs was observed between 1973 and 1990. This increasing trend is consistent with the temporal trend of concentrations in guillemot eggs (Sellström et al. 1990). In recent years, levels of PBDE have been increasing significantly in the Baltic Sea. Levels of PBDE in sediment from 1987 are about sixfold those of 1980 (total of PBDE congeners of Bromkal at 0.44 and 2.9 ng/g IG, respectively). PBDE levels are now of the same magnitude as PCB levels are (Nylund et al. 1992).

In sediments of Osaka Bay in Japan, tetra-, penta-, and hexa-BDE were found, and in 7 of 15 riverine and estuarine samples, deca-BDE was found in higher concentrations (Watanabe and Tatsukawa 1990). In mussels from Osaka Bay, only tetra-BDE was found (Watanabe et al. 1987). PBDEs were detected in the liver of cod from the North Sea. TBDE concentrations decreased from the southern to the northern part of the North Sea, but no temporal trend was observed between 1978 and 1987 (Fig. 2). 2,2',4,4'-tetra-BDE was identified in fish from a sewage pond in Schleswig Holstein, Germany (Wolf and Rimkus 1985). In herring from the North Sea, 2,2',4,4'-tetra-BDE was identified (de Boer 1990). In a pike from southwest Sweden sampled in 1981, PBDE was found by Andersson and Blomkist (1981) (Table 1). Sellström et al. (1993) confirmed the extremely high PBDE concentration in the same pike and perch samples (2–35 mg/kg lipid weight) and showed that it predominantly consisted of 2,2',4,4'-tetra-BDE. Other trout and bream samples from southern Sweden contained high concentrations of 2,2',4,4'-tetra-BDE as well. PBBs and PBDEs have also been detected in fish-eating birds and seals (Tables 1 and 2). The most frequently occurring PBB compound was hexa-BB. PBDEs were found in eels and livers of cormorants from Dutch freshwater areas (Rhine Delta and River Rur). With the exception of the River Rur, temporal decreases have been identified for PBDE concentrations in organisms from the Netherlands. The concentrations in eels from the River Rur have increased, possibly as a consequence of the use of PBDEs in German mining areas (de Boer 1990).

Table 1. PBDE concentrations in the environment (DW = dry weight; WW = wet weight; lipid = lipid basis).

Matrix	Location	Concentration (μg/kg)	Compound	References
Sediment	Osaka Bay (Japan)	11–30 DW	tetra-, penta-, hexa-BDE	Watanabe and Tatsukawa (1990)
–	Rivers (Japan)	33–375 DW	deca-BDE	Watanabe and Tatsukawa (1990)
Mussels	Osaka Bay (Japan)	15 WW	tetra-BDE	Watanabe et al. (1987)
Cod liver	Northern North Sea	26 lipid	tetra-BDE	de Boer (1989)
–	–	3 –	penta-BDE	de Boer (1989)
–	Central North Sea	54 –	tetra-BDE	de Boer (1989)
–	–	6 –	penta-BDE	de Boer (1989)
–	Southern North Sea	170 lipid	tetra-BDE	de Boer (1989)
–	–	22–26 lipid	penta-BDE	de Boer (1989)
Herring	Southern North Sea	100 lipid	tetra-BDE	de Boer (1990)
Eel	Rur River	1.4×10^3 lipid	tetra-BDE	de Boer (1990)
Pike	Southwest Sweden	27×10^3 lipid	PBDE[1]	Sellström et al. (1990)
Pike liver	Southwest Sweden	110×10^3 lipid	PBDE[1]	Sellström et al. (1990)
Eel	Southwest Sweden	17×10^3 lipid	PBDE[1]	Sellström et al. (1990)
Trout and bream	South Sweden	100–170 lipid	tetra-BDE	Sellström et al. (1993)
Cormorant liver	Rhine Delta	28×10^3 WW	PBDE	de Boer (1990)
Baltic Guillemot egg	Baltic Sea	2×10^3 lipid	PBDE[1]	Sellström et al. (1993)
Osprey	–	0.16–1.9×10^3 lipid	PBDE[1]	Sellström et al. (1993)
Ringed Seal	Northern Ice Sea	51 lipid	PBDE[1]	Sellström et al. (1993)
Baltic Grey Seal blubber	–	728 lipid	PBDE[1]	Sellström et al. (1993)
Seal blubber	Baltic Sea	730 lipid	PBDE	Andersson and Blomkist (1981)
–	–	26 lipid	tetra-BDE	Andersson and Blomkist (1981)

[1]Mainly tetra-BDE.

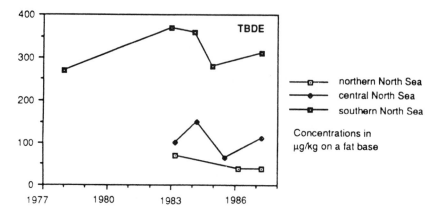

Fig. 2. Concentrations of 2,2′,4,4′-TBDE in cod liver from different regions of the North Sea (source: de Boer 1989).

IV. Toxicokinetics
A. Bioaccumulation

The degree of bioaccumulation is an important factor in the overall environmental risk of a compound. The accumulation potential of organic compounds has often been estimated successfully by measuring their partitioning coefficient in a two-phase solvent system of octanol and water (K_{ow}). n-Octanol is used as a model compound that should mimic the lipid

Table 2. Total (Σ-) PBB and PBDE concentrations calculated as technical mixture equivalents in herring, seals, and sea birds (μg/kg lipid) (Source: Jansson et al. 1987, 1993).

Organism	Area	Σ-PBB		Σ-PBDE	
		1987	1993	1987	1993
Herring	Baltic Sea		0.16		528
	Bothnian Gulf		0.09		123
	Skagerrak		0.27		735
Seal	Baltic Sea	20/26	90	728	
	Kattegat	3	10		
	Spitsbergen	4	40		
	Northern Ice Sea		0.42		51
Guillemot	Baltic Sea	160	370		
	North Sea		80		
	Northern Ice Sea	50	130		
Sea eagle	Baltic Sea	280		350	

Table 3. Some log K_{ow} values and log BCF (in Guppy, *Poecilia reticulata*) values for PBBs (Gobas 1989).

Compound	log K_{ow}	log BCF
4,4'-di-BB	5.72	5.43
2,4,6-tri-BB	6.03	5.06
2,2',5,5'-tetra-BB	6.50	6.16
2,2',4,4',6,6'-hexa-BB	7.20	5.85

pool in biota. Compounds with a high K_{ow} usually have a high affinity for animal lipids.

Accumulation and Excretion Data of PBBs. The log K_{ow} values and bio-concentration factors (log BCF) of some PBBs are given in Table 3. Gobas et al. (1989) reported BCF and K_{ow} values of polybrominated benzenes and biphenyls and could not detect a correlation between these values when K_{ow} exceeded 6. Possible reasons for an incorrect determination of BCF values are the elimination via feces or lower bioavailability due to absorption of the compounds to other molecules or to dissolved organic matter.

Zitko (1977) and Zitko and Hutzinger (1976) studied the bioaccumulation of PBBs in young Atlantic salmon (*Salmo salar*). The PBB was supplied through the water and food. To establish bioconcentration factors, a mixture of 388 µg PBB consisting of equal amounts of three di-bromo, one tri-bromo, and two tetra-bromobiphenyls was supplied to 3 L water to which the fish were exposed for 96 hr. The biomagnification factors were measured in an experiment where the fish were fed with a PBB-spiked diet. The same mixture of PBB compounds was used. The bromine content of the food was 7.75 µg g^{-1}. The reported bioconcentration and biomagnification factors are given in Table 4.

Table 4. Bioconcentration and biomagnification factors of PBBs on a wet weight basis in Salmon (*Salmo salar*) (Sources: Zitko and Hutzinger 1976; Zitko 1977).

Compound	Bioconcentration	Biomagnification
2,6-di-BB	1.2×10^6	0.179
2,4-di-BB	1.3×10^6	0.318
3,4-di-BB	63×10^3	0
2,5,4'-tri-BB	425×10^3	0.449
2,2',4,5'-tetra-BB	314×10^3	0.589
2,3',4',5-tetra-BB	110×10^3	0.571
C_6H_6	0	0
Firemaster BP-6	n.a.	1.00

For all PBBs, bioconcentration factors are three to four orders of magnitude higher than biomagnification factors. 3,4-di-BB was not accumulated at all from the food and hardly from the water. The corresponding 3,4-di-CB shows the same effect (Zitko 1977). No explanation was given for this phenomenon by the authors.

Bruggeman et al. (1982) showed for PCBs that solubility of the different congeners depends on chlorine substitution at the *ortho* position. Because 3,4-di-CB has no *ortho*-Cl, low solubility of the congeners may be the reason for low accumulation.

In Zitko's study (Zitko, 1977), compounds with a low bromine content bioconcentrated more strongly from water than compounds with a high bromine content. PBBs with more than six bromine atoms were hardly bioconcentrated at all.

However, from food, compounds with one to four bromine atoms were accumulated more when the bromine content was higher. PBBs with six to eight bromine atoms were only accumulated to a slight degree from food.

Zitko (1977) found hexa-BBs in fish tissue after exposure to a diet spiked with only octa-BBs. This dehalogenation is not known for higher chlorinated biphenyls. The accumulation of Firemaster from food appeared to be higher than the accumulation of other PBBs.

The half-life for excretion from fish was determined for two PBB congeners, Firemaster and OB (which is an octa-BB product) (Table 5).

Accumulation of PBDE. In Table 6, the log K_{ow} values of PBDEs are given. The accumulation of PBDEs with a low bromine content is greater than that of higher brominated compounds (Sellström et al. 1990). This is apparent from the often-used PBDE mixture Bromkal 70-5 DE, PBDE concentrations in sediments, and the PBDE pattern in herring and seals. Table 7 shows that the relative concentrations of tetra-BDE in seals is much higher than the tetra-BDE content in Bromkal and in sewage sludge.

Higher brominated diphenylethers are decreased in fish and fish consumers relative to the concentrations in the commercial products. This was not in agreement with what was found with the PCB congeners. The relative concentrations of PBDE congeners in herbivorous mammals in Sweden

Table 5. Biological half-lives of PBBs in fish, following uptake from food or water (Zitko 1977).

Compound	Uptake from water	Uptake from food
2,2',4,5' tetra-BB	21 d	28 d
2,4',5 tri-BB	13 d	26 d
Firemaster	n.a.	93 d
OB	n.a.	93 d

Table 6. Log K_{ow} values of PBDE
(Watanabe and Tatsukawa 1990).

PBDE	log K_{ow}
di-BDE	5.03
tri-BDE	5.47–5.58
tetra-BDE	5.87–6.16
penta-BDE	6.64–6.97
hexa-BDE	6.86–7.92
octa-BDE	8.35–8.90
deca-BDE	9.97

were the same as the concentrations in the commercial products (Jansson et al. 1993).

Comparison of accumulation of PBB and PBDE. Norris et al. (1974, 1975) studied the accumulation and excretion of deca-BDE and octa-BB in rats. The laboratory animals were fed for 2 yr with octa-BB or deca-BB 0.1 mg/kg fodder and during this period, at regular intervals, two animals were sacrificed. Deca-BDE caused an initial increase of bromine concentration in the liver, which did not increase further after 30 d. Bromine concentration in fat tissue increased only slightly. Octa-BB caused a rapid increase of bromine in the liver and in fat tissue; this increase was constant during the experiment. In an experiment with ^{14}C-octa-BB and ^{14}C-deca-BDE, the radioactivity of deca-BDE disappeared after 2 d through the feces. The ^{14}C of the octa-BB decreased to 33% during the first day. Later, the radioactivity decreased slowly, and after 16 d, 75% had disappeared. No toxic effects were found in the animals. The author concluded that deca-BDE does not accumulate, in contrast to octa-BB. The disadvantage of this study is that different numbers of bromine atoms were used. It may be that the difference in behavior is a consequence of the unequal bromine substitution and not of the difference between bromobiphenyl and bromodiphenylether.

Table 7. Percentages of PBDE congeners of Bromkal 70-5 DE (Sellström et al. 1990; Jansson et al. 1993).

	2,2',4,4'-tetra-BDE	penta-BDE (not defined)	2,2',4,4',5-penta-BDE
Bromkal 70-5 DE	44	8	48
Sewage sludge	40	9	51
Seal	89–92	3–5	2–6
Herring	62–80	6–11	9–21

No more general information is available about the differences between accumulation factors of PBBs and PBDEs. It is known that PCBs and PCBEs have accumulation factors of approximately the same magnitude (Zitko 1977). The validity of this relation is not confirmed for PBBs and PBDEs.

Comparison of accumulation of PBB, PBDE, and PCB.   PBBs and PCBs with one to four bromine or chlorine atoms have accumulation factors of approximately the same magnitude, but PBBs with more than five bromine atoms accumulate less than the corresponding PCBs. Higher chlorinated biphenyls are persistent in fish, while higher brominated biphenyls are partly debrominated (Zitko 1977; Zitko and Hutzinger 1976). On the other hand, 2,2',4,4',5,5'-He-BB was found to magnify by a factor of 140 in the food chain herring − grey seal (Jansson et al. 1993).

B. Interactions with Cytochrome P450

In a qualitative sense, the structure-effect relationships for BB congeners show a high similarity with those of chlorinated biphenyls (CBs). Quantitatively, bromine substitution appears to have a stronger effect than chlorine substitution on induction of the P450IA subfamily (3-MC type induction; Andres et al. 1983). This may be due to a higher affinity of BB congeners for the cytosolic Ah-receptor, because the same dose (150 μmol/kg, MW = 486) of BB-77 caused an eight times higher induction of ethoxyresorufin-O-deethylase (EROD) activity than CB-77, while the affinity for the Ah-receptor was also eight times higher.

2,3,7,8-Tetrachlorodibenzodioxin shows the highest affinity for the low capacity/high activity cytosolic receptor protein, which triggers all subsequent events; the activity of PBBs is three to four orders of magnitude lower.

Interactions between PCBs and the cytochrome P450-dependent monooxygenase (MO) systems were reviewed for fish-eating seabirds (Walker 1990) and marine mammals (pinnipeds, cetaceans, and the polar bear) (Boon et al. 1992).

The composition of Firemaster BP-6 and FF-1 is given in Table 8. The amount of Firemaster BP-6 produced is higher than that of Firemaster FF-1 (Safe 1984; Aust et al. 1987). BB-153 contributes more than 50% to both mixtures; it is a PB-type inducer, i.e., it induces the P450IIB subfamily. Mixed-type inducers present in BP-6 are BB-118 (2.9%) and BB-167 (7.95%) (Robertson et al. 1980). The pure 3-MC type inducers BB-77, BB-126, and BB-169 contribute 0.16%, 0.08%, and 0.29% to this mixture, respectively. The contributions of individual congeners of BBs to BP-6 are very different from the contribution of CB congeners to any of the Aroclor or Clophen mixtures (Schulz et al. 1989). It has been shown (Aust et al. 1987) that the BB congeners of BP-6 are entirely responsible for induction

Table 8. Composition of Firemaster BP-6 and Firemaster FF-1 (Source: Silberhorn et al. 1990).Systematic numbering of individual congeners after the system developed for PCBs by Ballschmiter and Zell (1980).

No.	Structure	BP-6 (%)	FF-1 (%)
Di-BB	4.4'	<0.020	–
15			
Tri-BB	2,2',5-	0.050	–
18			
26	2,3',5-	0.024	–
31	2,4',5-	0.015	–
37	3,4',4-	0.021	–
Tetra-BB	2,2',4,5'-	0.025	–
49			
52	2,2',5,5'-	0.052	–
53	2,2',5,6'-	<0.013	–
66	2,3',4,4'-	0.028	–
70	2,3',4',5-	0.017	–
77	3,3',4,4'-	0.159	0.03
Penta-BB	2,2',3,5',6-	–	0.02
95			
99	2,2',4,4',5-	–	0.03
101	2,2',4,5,5'-	2.69	1.54
114	2,3,4,4',5-	0.08	–
118	2,3',4,4',5-	2.94	0.8
126	3,3',4,4',5-	0.079	<0.01
Hexa-BB	2,2',3,4,4',5'-	12.3	5.23
138			
141	2,2',3,4,5,5'-	0.10	–
149	2,2',3,4',5',6-	2.24	0.78
153	2,2',4,4',5,5'-	53.9	55.2
156	2,3,3',4,4',5-	0.98	0.37
157	2,3,3',4,4',5'-	0.526	0.05
167	2,3',4,4',5,5-	7.95	3.37
168	2,3',4,4',5',6-	<0.025	–
169	3,3',4,4',5,5'-	0.294	0.15
Hepta-BB	2,2',3,3',4,4'5-	0.256	1.66
170			
172	2,2',3,3',4,5,5'-	–	0.15
174	2,2',3,3',4,5,6'-	–	0.24
180	2,2',3,4,4',5,5'-	6.97	23.5
187	2,2',3,4',5,5',6-	0.392	–
189	2,3,3',4,4',5,5'-	–	0.51
Octa-BB	2,2',3,3',4,4',5,'-	–	1.65
194			
196	2,2',3,3',4,4',5',6-	–	0.31
203	2,2',3,4.4',5,5',6-	–	0.30
Total		92.09	95.89
Approximate molecular weight		628	650

of the MO system in rats; thus, there is no additional effect of impurities, such as brominated dibenzofurans in the mixture.

Another mixture of PBBs is Firemaster FF1. Next to BB-153, BB-180 dominates in this mixture, which is also a PB-type inducer. In general, the 3-MC or mixed-type inducers show a somewhat lower contribution to this mixture compared with BP-6, with the exception of BB-170.

P450 induction occurs at lower body burdens in infant compared to adult female rats: when 1 mg kg^{-1} was injected during a period of 18 d into the mother animals, it caused (a mixed function) MO induction in the sucklings but not in the mothers (Aust et al. 1987).

Carlson (1980a,b) investigated the inducing potential of PBDEs with a mixture of a low overall bromination (24% tetra-BDE congeners, MW = 502; and 50% penta-BDE congeners, MW = 582), a higher overall bromination (45% hepta-BDE congeners, MW = 662; and 30% octa-BDE congeners, MW = 822), and the deca-BBE congener only (BB-209, MW = 982). Both mixtures induced O-ethyl O-p-nitrophenyl phenylphosphothionate and Uridine Diphosphate-glucuronyltransferase, which catalyzes the conjugation of hydroxylated compounds to glucuronic acid (phase II metabolism). p-Nitroanisole demethylase and arylhydrocarbon hydroxylase were induced most by the mixture of lowest bromination. The deca-BBE congener BB-209 did not cause any enzyme induction. A long-term study where rats were injected with a daily dose of 0.8–3 μmol/kg of both mixures for 90 d showed that the abovementioned enzymes were induced at a concentration of as low as 1 μmol/kg. Moreover, the enzymes remained induced even 30–60 d after the termination of exposure. These results demonstrate that these inducers are not only potent but that their effects may be long-lasting.

C. Biotransformation of PBBs

In vitro metabolism and structure-effect relationships.   Aust et al. (1987) concluded that BBs with H atoms at an *ortho-* and *meta-* (*o,m*) position, and a maximum of four bromine atoms, were metabolized by P450IA in microsomal systems of rats. Mills et al. (1985) drew similar conclusions, with the remark that bromine substitution would be necessary at least at one *meta-* or *para-*position for metabolism to occur because 2,2′-di-BB was not metabolized at all, whereas 2,2′,4,4′-tetra-BB was metabolized. Rates of metabolism decreased in the order 4,4′-di-BB > 3,4,4′-tri-BB > 3,3′,4,4′-tetra-BB > 2,3,3′4′-tetra-BB > 2,3′,4′,5-tetra-BB > 2,2′,4,5′-tetra-BB. Thus, metabolism by microsomes from 3-MC-induced rats decreased with increasing numbers of *ortho*-Br atoms and thus with an increasing energy barrier for a planar configuration. Such quantitative structure-activity relationships for PBBs are not available for marine organisms.

In the case of PCB congeners with vicinal H atoms in the *o,m* positions, they appear to be always capable of being metabolized in the polar bear, and metabolized in seals and cetaceans only when a maximum of one *or-*

tho-Cl is present, and persistent in seabirds (Walker 1990; Boon et al. 1992).

In contrast to the situation for 3-MC-induced rats, 2,2'-di-BB was metabolized very rapidly by microsomes from PB-induced rats. According to Mills et al. (1985) and Aust et al. (1987), congeners with vicinal H atoms at *m,p* positions are metabolized by P450IIB. The rates decreased in the order: 2,2'-di-BB > 2,2',4,5'-tetra-BB > 2,3',4',5-tetra-BB > 2,2', 4,5,5'-penta-BB.

In the case of PCBs in marine animals, congeners with vicinal H atoms in the *m,p* positions appear capable of being metabolized by seabirds, seals, and the polar bear, but much less so by cetaceans, which lack P450IIB enzymes.

Because different groups of marine animals show such a varying scope for metabolism of PCBs, it is highly unlikely that the structure-effect relationships discussed above for the biotransformation of PBBs in rats also represent the situation in marine animals. Studies of these processes are urgently needed.

In vivo metabolism. Metabolites of congeners from Firemaster were detected in pigs (a hydroxylated penta-BB) and dogs (2,2',4,4',5,5'-hexabromobiphenylol) (Safe 1984). Zitko and Hutzinger (1976) reported a dibromodiphenylol in salmon. Of the more extensively studied PCBs, hydroxylated as well as methylsulfone metabolites have been reported (Jansson et al. 1987; Safe 1984). The metabolites are possibly formed via an epoxide, but this is not necessarily the first step (Bush and Trager 1985; Trager 1989) leading either to a hydroxylated compound (phenolic metabolite) or to binding of a reactive intermediate (epoxide) to glutathione (GSH).

Millis et al. (1985) compared the toxicokinetics of the P450IA inducers 3,3',4,4'-tetra-BB (BB-77, MW = 486) and 3,3',4,4',5,5'-hexa-BB (BB-169, MW = 646). A single dose of 21.3 μmol/kg of a congener was administered to one of two groups of rats. The concentration of BB-169 in liver and adipose tissue and the activity of aryl hydrocarbon hydroxylase (AHH) reached a maximum 1 d after injection and remained at this maximum throughout the rest of the experiment. In contrast, the levels of BB-77 and AHH activity declined sharply after 2 d in the other group of rats. The decline of BB-77 was attributed to metabolism; in contrast to BB-169, BB-77 possesses vicinal H atoms. At the end of the experiment, livers of animals injected with BB-169 showed histological changes similar to tetrachlorodibenzo-*p*-dioxin (TCDD) effects, but the livers of BB-77 injected animals did not, even though the affinity for the TCDD receptor protein was 10 times higher for BB-77 than for BB-169. It may be concluded that metabolism of BB-77 decreased its TCDD-type toxicity.

The rate of *in vitro* metabolism of BB-77 is inhibited by the addition of BB-169. At equal concentrations (3 μM), the metabolism of BB-77 is almost completely inhibited (Mills et al. 1985) because BB-169 binds to the enzyme receptors and cannot be metabolized and because vicinal H atoms are lacking. BB-169 induces both P450IA1 and IA2, but only binds to IA2 (Voorman and Aust 1987). The BB-169/protein complex appeared to be more stable than P450IA2 by itself. The binding between BB-169 and P450IA2 was noncovalent and could be broken down by extraction with dichloromethane. BB-169 also inhibits the estradiol-2-hydroxylase activity of purified P450IA2 in a noncompetitive way (Voorman and Aust 1988).

V. Toxicity of PBBs
A. Acute Toxicity

The acute toxicities of a number of PBB and PCB mixtures are summarized in Table 9. Firemaster BP-6 appears to have a similar acute toxicity to rats as the PCB mixtures Aroclor 1254 and Kanechlor 500.

Gupta et al. (1983b) administered a dose of 100 mg Firemaster BP-6 kg^{-1} to rats on 22 occasions over a period of 1 mon. After 90 d, all females had died, but 62% of the male rats still survived. In a similar experiment with a daily dose of 30 mg kg^{-1}, all animals of both sexes survived for 90 d.

B. Toxic Effects in Relation to Cytochrome P450 Induction

The toxicity of BB congeners strongly depends on their molecular structure. Induction of the P450IA subfamily of P450 is the precursor of a whole spectrum of possible effects at more integrated levels of biological structure: weight loss, thymus atrophy, and changes in the liver, such as proliferation of the smooth endoplasmatic reticulum (location of the P450 system), increased RNA and protein content, decreased DNA content, cell necrosis, liver enlargement, and hepatic porphyria (Render et al. 1982; Jensen et al. 1983; Koster et al. 1980).

Table 9. Acute toxicities of PBB and PCB mixtures. AHH = aryl hydrocarbon hydroxylase activity (Sources: [1]Gupta et al. 1983a; [2]Andres et al. 1983; [3]Safe 1984).

Mixture	Species/Sex	Details	$LD_{50}/EC_{50}/LC_{50}$
Firemaster BP-6 (PBB)	Rat (F)		LD_{50}: 65 mg/kg/d[1]
Firemaster BP-6 (PBB)	Rat (M)		LD_{50}: 149 mg/kg/d[1]
Firemaster BP-6 (PBB)	Rat	AHH activity	EC_{50}: 50–55 mg/kg[2]
Aroclor 1254 (PCB)	Rat	AHH activity	EC_{50}: 50–55 mg/kg[2]
Kanechlor 500 (PCB)	Rat	AHH activity	EC_{50}: 50–55 mg/kg[2]
Firemaster BP-6 (PBB)	Rat	Oral	LD_{50}: 21.5 g/kg[3]
Hexa-BB	Rabbit	Skin	LD_{50}: 5 g/kg[3]
Firemaster FF 1 (PBB)	Mink	Food	LC_{50}: 3.95 mg/kg[3]

C. PBBs and Cancer

The carcinogenicity of PBBs and PCBs has been reviewed by Silberhorn et al. (1990). The formation of tumors is a multistage process. The first phase is an irreversible mutation of DNA, the initiation phase. The growth of an initiated cell to a tumor is the promotion phase. There are strong indications that PBBs, PCBs, dioxins, dibenzofurans, and related compounds are not mutagenic compounds but do promote the carcinogenicity of mutagenic compounds, such as nitrosamine and certain polyaromatic hydrocarbons (PAHs) (Safe 1984; Jensen et al. 1983; Kavanagh et al. 1985). The latter may well be of importance for the marine environment because halogenated biphenyls often co-occur with PAHs derived from oil or combustion processes. Tumor promotion has been reported for 3-MC type (BB-169; Jensen et al. 1983; Kavanagh et al. 1985) as well as PB-type inducers (BB-153; Kavanagh et al. 1985). The tumor-promoting capacity of Firemaster BP-6 is greater than that of its dominant congener, BB-153 (Jensen et al. 1982).

Safe (1984) and Aust (1987) reported liver cancer in rats from PBBs without the addition of an initiator, but the single doses of 200 mg kg^{-1} and 1 g kg^{-1}, respectively, were high. Initiating compounds may have been present in the diet, because in the second study, for example, neoplastic nodules were observed after 6 mon, which developed into malignant tumors after a period of 2 yr. (Jensen and Sleight 1986).

D. Dermal Toxicity

Like PCBs, PBBs cause chloracne; in monkeys, this occurred at a concentration of 50 mg kg^{-1} in the diet (Safe 1984). After 20 wk, exposure at a dose of 2 mg/animal twice weekly, Firemaster FF-1 also caused skin papillomas in previously initiated mice (Poland et al. 1982).

E. Neurotoxicity

Firemaster BP-6 caused chronic and subchronic neurotic symptoms, such as irritation, changed behavior and decreased muscular control (Safe 1984).

F. Immunotoxicity

PBBs caused thymus atrophy, hypersensitivity, decreased antibody response, and decreased resistance against infections in guinea pigs (Safe 1984). However, the mechanism remained unclear. A decreased immune response in rats and mice occurred after exposure to Firemaster BP-6 at either 30 mg kg^{-1} for 22 d or 10 mg kg^{-1} for 6 mon. Doses of 0.1–100 mg kg^{-1} for 30 d resulted in a reduction of B- and T-helper cells (Aust et al. 1987). In general, immunosuppression by PBBs occurs at levels that also cause a number of the other toxic effects described.

G. Effects on Reproduction and the Regulation
of Steroid Hormones

Because the placenta is not an efficient barrier to PBBs and related lipo-philic compounds, these enter the developing embryo during pregnancy. A second source after delivery is a mother's milk (Safe 1984). Over a period of 7 mon, a dietary concentration of 0.3 mg kg^{-1} of Firemaster FF-1 caused a longer sexual cycle and decreased progesterone concentrations in monkeys, and weights of neonates decreased. A dose of 40 and 200 mg/kg Firemaster BP-6 administered to rats once during pregnancy caused no teratogenic effects. Malformations of fetuses occurred at single doses of 400 and 800 mg/kg. The 800 mg/kg doses resulted in much higher fre-quency than at 400 mg/kg (Beaudoin, 1977).

A dietary concentration of 100 mg PBBs kg^{-1} caused decreased egg production and nesting behavior in Japanese quail (Aust et al. 1987). New-ton et al. (1982) reported increased hydroxylation rates of testosterone to 7α- and 6β-hydroxytestosterone in rats fed 100 mg kg^{-1} PBB for 4 mon. A reduction of testosterone to dihydrotestosterone and dihydroandrosterone also was found in microsomes from both sexes.

Byrne et al. (1988) fed rats for 5–7 mon with doses of 1–50 mg kg^{-1} food. Lowered concentrations of serum corticoid hormones were already observed at low doses of Firemaster BP-6, together with a decrease in weight of the adrenals.

H. Influence on Vitamin A and Thyroid Hormone Regulation

A single dose of 2 mg kg^{-1} of BB-169 caused an increase by a factor of 2 in vitamin A breakdown products in urine and feces (Cullum and Zile 1985).

Gupta et al. (1983a) found decreased serum TT4 and triiodothyronine concentrations after exposure over 125 d to concentrations of 0.1–10 mg Firemaster BP-6 kg^{-1}.

VI. Toxicity of PBDEs

Most toxicological studies concerning PBDEs are conducted with a com-mercial BDE composition consisting of 77.4% deca-BDE, 21.8% nona-BDE, and 0.8% octa-BDEs (Norris 1974; Norris 1975; National Toxicology Program (NTP) 1986). Toxicity data of this BDE-mixure do not reflect the toxicity of all BDE congeners. Absorption of this mixture from the gastrointestinal tract is low (approximately 1% in rats) and toxicity will be low compared to that of other BDE congeners (see below).

A. Acute Toxicity

Intragastric intubation of single doses of up to 2000 mg kg^{-1} of commercial BDE to female rats did not affect growth rate, and no gross pathological effects were encountered during a 14-d post-treatment observation period (Norris et al. 1975).

In feeding studies, rats and mice of both sexes were exposed to a range of 5–100 g kg^{-1} of commercial BDE in the diet during 14 d and 13 wk, respectively. There were no effects on survival, body weight, or food consumption, and no gross or microscopic pathological effects (NTP 1986).

In another study with rats (Norris et al. 1975), diet concentrations were 0.1–10 g kg^{-1} commercial BDE during 30 d. Concentrations of 1 and 10 g kg^{-1} caused increased liver weight and a dose-related thyroid hyperplasia. Histopathological investigations revealed liver and kidney lesions at the highest concentration. A concentration of 0.1 mg kg^{-1} commercial BDE in food, corresponding to 8 mg kg^{-1}, body weight, is considered as an unequivocal no-effect level.

B. Toxic Effects in Relation to Cytochrome P450 Induction

As is the case with BB congeners, the toxicity of BDE congeners strongly depends on their molecular structure. BDEs induce the same isoenzymes of cytochrome P450 as PBBs and PCBs. Possible effects are described in section V B.

A dose of 0.1 mMol kg^{-1} d^{-1} BDE (main components: 24.6% tetra-BDEs, 58.1% penta-BDEs, and 13.3% hexa-BDEs) administered to male rats during 14 d increased cytochrome P450 more than an equal molar dose of a BDE mix consisting of 45.1% hepta-BDEs, 30.7% octa-BDEs, and 13.0% nona-BDE; an equal molar dose of high-purity deca-BDE did not significantly increase cytochrome P450. Penta-BDE increased liver/body weight with 64%, octa-BDE with 45%, and deca-BDE with 25% (Carlson 1980a).

C. Carcinogenicity

In feeding studies, female and male rats and mice were exposed to 25 and 50 g kg^{-1} commercial BDE in the diet during 103 wk, resulting in increased incidence of neoplastic nodules in livers of male and female rats. There was equivocal evidence of carcinogenicity for male mice as shown by increased incidences of hepatocellular adenomas and carcinomas. However, these were not increased in comparison with control groups of earlier experiments. Incidences of follicular adenomas or carcinomas of the thyroid gland were only marginally increased. There was no evidence of carcinogenicity for female mice receiving commercial BDE. Effects observed in these studies were attributed to the less brominated BDEs of the commercial mixture (NTP 1986).

D. Mutagenicity

Commercial BDE was not mutagenic in bacteria, *Salmonella typhimurium*, or in a mouse lymphoma assay. It did not induce chromosomal effects like sister-chromatid exchanges or chromosomal aberrations in Chinese hamster ovary cells *in vitro* (NTP 1986).

E. Dermal Toxicity

Commercial BDE did not irritate the skin of rabbits and rats and was only mildly irritating to the eyes of rabbits. External exposure caused no chloracne of the ear of rabbits (Norris et al. 1975).

F. Effects on Reproduction, Embryotoxicity, and Teratogenicity

In sticklebacks, *Gasterosteus aculeatus*, Holm et al. (1993) found a decrease in spawning success. Female fish received a dietary dose of approximately 0.5 mg Bromkal (for composition, see Table 7) in 3.5 mon (concentration in the food 346 mg kg^{-1}, with fish consuming 2% of their body weight/d). Uptake efficiency was approximately 20%. Only 2 of 10 fish spawned, compared with 8 of 10 in the controls.

Commercial BDE did not affect the reproductive capacity of rats. Concentrations in the diet of 3, 30, or 100 mg kg^{-1} given 90 d before mating as well as during mating, gestation, and lactation had no effects on the number of pregnancies or on survival and weight of the neonates (Norris et al. 1974).

Oral administration of 2–15 mg kg^{-1} body weight per d^{-1} of a PBDE mixture with a lower overall degree of bromination (0.2% penta-BDEs, 8.6% hexa-BDEs, 45% hepta-BDEs, 33.5% octa-BDEs, 11.2% nona-BDE, and 1.4% deca-BDE) on gestation days 7–19 to pregnant rabbits did not cause teratogenic responses. At 15 mg kg^{-1} slight fetal toxicity was observed by an increase in the incidence of delayed ossification of the breast bone (Breslin et al. 1989).

Commercial BDE caused no teratogenic response in fetuses of rats intubated with 10–1000 mg kg^{-1} d^{-1} on gestation days 6–15. Fetal toxicity only occurred at 1000 mg kg^{-1} as subcutaneous edema and a delayed ossification of normally developed bones of the fetal skull (Norris et al. 1975).

VII. Risk Evaluation and Recommendations

Both classes of brominated fire retardants have properties similar to those of related compounds, such as the PCBs, dibenzodioxins and dibenzofurans, and the PAHs. Thus, their toxicity is likely to interfere with the toxicity of these related compounds because the important mechanisms of toxicity of PBBs and PBDEs are also shown by them.

The scientific basis for a risk evaluation of PBBs and especially the PBDEs in the aquatic environment is very small. To improve this, more knowledge is required, especially in the following areas:

1) their actual concentrations in different compartments of the marine environment, especially in sediments and different classes of biota. As for all commercial mixtures of compounds with the same basic structure, only the quantification of individual congeners will allow comparative environmental studies. Because none of the individual BDE congeners and only

some BB congeners are yet available as analytical standards, there is an urgent need for more.

2) the toxicity mechanisms of the individual congeners in marine animals, including toxicokinetic aspects, in order to gain insight into the toxicity of congener mixtures of the same class and congeners of the different classes of related compounds.

Because of the scarcity of data, it may be only tentatively concluded that the present concentrations of PBBs and PBDEs in marine food chains of the Baltic Sea, the North Sea, and the North Atlantic Ocean do not by themselves yet appear to represent a great environmental risk.

Despite this fact, their replacement by environmentally less harmful alternatives is strongly recommended because they are often used in open systems. Much of the quantities of PBB and PBDEs produced will thus eventually reach the marine environment, where they are likely to accumulate because of their resistance to degradative processes. Finally, because of their environmental properties, the continued release of PBBs and PBDEs represents an increasing risk to marine organisms.

Summary

Data on two classes of brominated polyaromatic flame retardants are reviewed with emphasis on analytical aspects, occurrence, fate, and toxicity in the environment. Concentrations of brominated fire retardants are quantified as equivalents of commercial mixtures. Because different congeners behave differently in the environment and show large differences in toxicity, future studies would benefit from the availability of analytical standards of individual congeners.

The main environmental properties and mechanisms of toxicity of the PBBs and PBDEs are similar to those of the structurally related PCBs and dibenzodioxins. Although the present concentrations of brominated fire redardants do not yet appear to represent a major environmental risk in marine food chains, their replacement by environmentally less harmful alternatives is recommended.

References

Andersson Ö, Blomkist G (1981) Polybrominated aromatic pollutants found in fish in Sweden. Chemosphere 10:1051–1060.
Andres J, Lambert I, Robertson L, Bandiera S, Sawyer T, Lovering S, Safe S (1983) The comparative biologic and toxic potencies of polychlorinated biphenyls and polybrominated biphenyls. Toxicol Appl Pharmacol 70:204–215.
Arias P (1992) Brominated diphenyloxides as flame retardants. Bromine based chemicals. Draft report of OECD (Organization for Economic Cooperation and Development). Brussels, October 1992.
Aust SD, Millis CD, Holcomb L (1987) Relationship of basic research in toxicology

to environmental standard setting: the case of polybrominated biphenyls in Michigan. Arch Toxicol 60:229–237.

Ballschmiter K, Zell M (1980) Analysis of polychlorinated biphenyls (PCB) by glass capillary gas chromatography. Composition of technical Aroclor and Clophen-PCB mixtures. Fresenius J Anal Chem 302:20–31.

Beaudoin AR (1977) Teratogenicity of polybrominated biphenyls in rats. Environ Res 14:81–86.

Boon JP, van Arnhem E, Jansen S, Kannan N, Petrick G, Schulz D, Duinker JC, Reijnders PJH, Goksøyr A (1992) The toxicokinetics of PCBs in marine mammals with special reference to possible interactions of individual congeners with the cytochrome P450-dependent monooxygenase system. An overview. In: Walker CH, Livingstone D (eds) Persistent pollutants in marine ecosystems. Lewis Publishers, Chelsea, MI, pp 119–159.

Breslin WJ, Kirk HD, Zimmer MA (1989) Teratogenic evaluation of a polybromodiphenyl oxide mixture in New Zealand white rabbits following oral exposure. Fund Appl Toxicol 12:1151–1157.

Bruggeman WA, van der Steen J, Hutzinger O (1982) Reversed-phase thin-layer chromatography of polynuclear aromatic hydrocarbons and chlorinated biphenyls. Relationship with hydrophobicity as measured by aqueous solubility and octanol-water partition coefficient. J Chromatogr 238:335–346.

Buser HR (1986) Polybrominated dibenzofurans and dibenzo-p-dioxins: thermal reaction products of polybrominated diphenyl ether flame retardants. Environ Sci Technol 20:404–408.

Bush ED, Trager WF (1985) Substrate probes for the mechanism of aromatic hydroxylation catalysed by cytochrome P450 selectively deuterated analogs of warfarin. J Med Chem 28:992–996.

Byrne JJ, Carbone JP, Pepe MG (1988) Suppression of serum adrenal cortex hormones by chronic low-dose polychlorobiphenyl or polybromobiphenyl treatments. Arch Environ Contam Toxicol 17:47–53.

Carlson GP (1980a) Induction of xenobiotic metabolism in rats by short-term administration of brominated diphenylethers. Toxicol Lett 5:19–25.

Carlson GP (1980b) Induction of xenobiotic metabolism in rats by polybrominated diphenyl ethers administered for 90 days. Toxicol Lett 6:207–212.

Cullum ME, Zile MH (1985) Acute polybrominated biphenyl toxicosis alters Vitamin A homeostasis and enhances degradation of Vitamin A. Toxicol Appl Pharmacol 81:177–181.

de Boer J (1988a) Trends in chlorobiphenyl contents in livers of Atlantic cod (Gadus morhua) from the North Sea, 1979–1987. Chemosphere 17:1811–1819.

de Boer J (1988b) Chlorobiphenyls in bound and non-bound lipids of fishes: comparison of different extraction methods. Chemosphere 17:1803–1810.

de Boer J (1989) Organochlorine compounds and bromodiphenylethers in livers of Atlantic cod (Gadus morhua) from the North Sea, 1977–1987. Chemosphere 18:2131–2140.

de Boer J (1990) Brominated diphenyl ethers in Dutch freshwater and marine fish. In: Hutzinger O, Fiedler H (eds) Proceedings of the 10th International Symposium on Dioxins '90, Bayreuth, Germany 2:315–318.

de Kok JJ, de Kok A, Brinkman UAT (1977) Analysis of polybrominated biphenyls by gas chromatography with electron capture detection. J Chromatogr 143:367–383.

Domino EF, Wright DD, Domino SE (1980) GCEC analysis of polybrominated biphenyl constituents of Firemaster FF-1 using tetrabromobiphenyl as an internal standard. J Anal Toxicol 4:299–304.

Erickson MD, Kelner L, Bursey JT, Rosenthal D, Zweidinger RA, Pelizzari ED (1980) Method for the analysis of brominated biphenyls by gas chromatography mass spectrometry. Biomed Mass Spect 7:99–104.

Farrell TJ (1980) Glass capillary gas chromatography of chlorinated dibenzofurans, chlorinated anisoles and brominated biphenyls. J Chromatogr Sci 18:10–17.

Fehringer NV (1975) Determination of polybrominated biphenyl residues in dry animal feed. J Assoc Off Anal Chem 58:1206–1210.

Gobas FAPC, Clark KE, Shiu WY, Mackay D (1989) Bioconcentration of polybrominated benzenes and biphenyls and related superhydrophobic chemicals in fish: role of bioavailability and elimination into the feces. Environ Toxicol Chem 8:231–245.

Gupta BN, McConnell EE, Goldstein JA, Harris MW, Moore JA (1983a) Effect of a polybrominated biphenyl mixture in the rat and mouse: 1. Six month exposure. Toxicol Appl Pharmacol 68:1–18.

Gupta BN, McConnell EE, Goldstein JA, Harris, MW, Moore JA (1983b) Effect of a polybrominated biphenyl mixture in the rat and mouse: 2. Lifetime study. Toxicol Appl Pharmacol 68:19–35.

Holm G, Norrgren L, Andersson T, Thurén A (1993) Effects of exposure to food contaminated with PBDE, PCN or PCB on reproduction, liver morphology and cytochrome P450 activity in the three-spined stickleback, *Gasterosteus aculeatus*. Aquat Toxicol 27:33–50.

Jansson B, Asplund L, Olsson M (1987) Brominated flame retardants — Ubiquitous environmental pollutants. Chemosphere 16:2353–2359.

Jansson B, Andersson R, Asplund L, Bergman A, Litzen K, Nylund K, Reutergardh L, Sellström U, Uvemo UB, Wahlberg C, Widequist U (1991) Multiresidue method for the gas chromatographic analysis of some polychlorinated and polybrominated pollutants in biological samples. Fresenius J Anal Chem 340:439–445.

Jansson B, Andersson R, Asplund L, Litzen K, Nylund K, Sellstrom U, Uvemo U, Wahlberg C, Wideqvist U, Odsjo T, Olsson M (1993) Chlorinated and brominated persistent organic samples from the environment. Environ Toxicol Chem 12:1163–1174.

Jensen RK, Sleight SD, Goodman JI, Aust SD, Trosko JE (1982) Polybrominated biphenyls as promotors in experimental hepatocarcinogenesis in rats. Carcinogenesis 3:1183–1186.

Jensen RK, Sleight SD, Aust SD, Goodman JI, Trosko JE (1983) Hepatic tumor-promoting ability of 3,3′,4,4′,5,5′-hexabromobiphenyl: the interrelationship between toxicity, induction of hepatic microsomal drug metabolizing enzymes, and tumor-promoting ability. Toxicol Appl Pharmacol 71:163–176.

Jensen RK, Sleight SD (1986) Sequential study on the synergistic effects of 2,2′,4,4′,5,5′-hexabromobiphenyl and 3,3′,4,4′,5,5′-hexabromobiphenyl on hepatic tumor promotion. Carcinogenesis 7:1771–1774.

Kavanagh TJ, Rubenstein C, Liu PL, Chang CC, Trosko JE, Sleight SD (1985) Failure to induce mutation in Chinese hamster V79-cells and WB red liver cells by the polybrominated biphenyls Firemaster BP-6, 2,2',4,4',5,5'-hexabromobiphenyl, 3,3',4,4',5,5'-hexabromobiphenyl and 3,3',4,4'-tetrabromobiphenyl. Toxicol Appl Pharmacol 79:91–98.

Koster P, Debets FMH, Strik JJTWA (1980) Porphyrinogenic action of fire retardants. Bull Environ Contam Toxicol 25:313–315.

Millis CD, Mills RA, Sleight SD, Aust SD (1985) Toxicity of 3,4,5,3',4',5'-hexabrominated biphenyl and 3,4,3',4'-tetrabrominated biphenyl. Toxicol Appl Pharmacol 78:88–95.

Mills RA, Millis CD, Dannan GA, Guengerich FP, Aust SD (1985) Studies on the structure-activity relationships for the metabolism of polybrominated biphenyls by rat liver microsomes. Toxicol Appl Pharmacol 78:96–104.

Morris PJ, Quensen JF, Tiedje JM, Boud SA (1993) An assessment of the reductive debromination of polybrominated biphenyls in the Pine River reservoir. Environ Sci Technol 27:1580–1586.

Mulligan KJ, Caruso JA (1980) Determination of polybromobiphenyl and related compounds by gas-liquid chromatography with a plasma emission detector. Analyst 105:1060–1067.

National Toxicology Program (NTP) (1986) Toxicology and carcinogenesis studies of decabromodiphenyl oxide in F344/N rats and B6C3F1 mice (feed studies). Technical report series no. 309, NIH publication No. 86-2565, U.S. Department of Health and Human Services, Washington, DC.

Newton JF, Braselton Jr WE, Lepper LF, McCormack KM, Hook JB (1982) Effects of polybrominated biphenyls on metabolism of testosterone by rat hepatic microsomes. Toxicol Appl Pharmacol 63:142–149.

Norris JM, Ehrmantraut JW, Gibbons CL, Kociba RJ, Schwets BA, Rose JQ (1974) Toxicological and environmental factors involved in the selection of decabromodiphenyl oxide as a fire retardant chemical. J Fire Flam Combust Toxicol 1:52–77.

Norris JM, Kociba RJ, Schwets BA, Rose JQ, Humiston CG, Jewett GL, Gehring PJ, Mailhes JB (1975) Toxicology of octabromobiphenyl and decabromodiphenyl oxide. Environ Hlth Persp 11:153–161.

Nylund K, Asplund L, Jansson B, Jonsson P, Litzén K, Sellström U (1992) Analysis of some polyhalogenated organic pollutants in sediment and sewage sludge. Chemosphere 24:1721–1730.

Poland A, Palen D, Glover E (1982) Tumor promotion by TCDD in skins of HRS/J hairless mice. Nature 300:271–273.

Render JA, Aust SD, Sleight SD (1982) Acute pathologic effects of 3,3',4,4',5,5'-hexabromobiphenyl in rats: comparison of its effects with Firemaster BP-6 and 2,2',4,4',5,5'-hexabromobiphenyl. Toxicol Appl Pharmacol 62:428–444.

Robertson LW, Parkinson A, Safe S (1980) Induction of both cytochromes P-450 and P-448 by 2,3',4,4',5-pentabromobiphenyl, a component of Firemaster. Biochem Res Commun 92:175–182.

Ruzo LO, Sündstrom G, Hutzinger O, Safe S (1976) Photodegradation of polybromobiphenyls (PBB). J Agric Food Chem 24:1062–1065.

Safe S (1984) Polychlorinated biphenyls (PCBs) and polybrominated biphenyls (PBBs): Biochemistry, toxicology, and mechanism of action. CRC Crit Rev Toxicol 13:319–395.

Schulz DE, Petrick G, Duinker JC (1989) Complete characterization of polychlorinated biphenyl congeners in commercial Aroclor and Clophen mixtures by multidimensional gas chromatography-electron capture detection. Environ Sci Technol 23:852–859.

Sellström U, Jansson B, Nylund K, Odsjö T, Olsson M (1990) Anthropogenic brominated aromatics in the Swedish environment. Dioxin 1990 EPRI-Seminar. Bayreuth, Germany. Short papers, pp 357–360.

Sellström U, Jansson B, Kierkegaard A, de Wit C (1993) Polybrominated diphenylethers (PBDE) in biological samples from the Swedish environment. Chemosphere 26:1703–1718.

Silberhorn EM, Glauert HP, Robertson LW (1990) Carcinogenicity of halogenated biphenyls PCBs and PBBs. Crit Rev Toxicol 20:440–496.

Svensson S, Hellsten E (1989) Flame retardants – Production and use in Sweden. Proceedings Workshop on Brominated Flame Retardants, Skokloster, Sweden, 24–26 October 1989.

Sweetman JA, Boettner EA (1982) Analysis of polybrominated biphenyls by gas chromatography with electron capture detection. J Chromatogr 236:127–136.

Tanabe S (1988) PCB problems in the future: foresight from current knowledge. Environ Pollut 50:5–28.

Trager WF (1989) Isotope effects as mechanistic probes of cytochrome P450 catalyzed reactions. In: Baillie TA, Jones JR (eds) Proceedings of International Symposium on Synth Appl Isot Labelled Cpd 1988, Seattle, WA, pp 333–340.

van der Valk F, Dao QT (1988) Degradation of PCBs and HCB from sewage sludge during alkaline saponification. Chemosphere 17:1735–1739.

Voorman R, Aust SD (1987) Specific binding of polyhalogenated aromatic hydrocarbon inducers of cytochrome P-450 d to the cytochrome and inhibition of its estradiol 2-Hydroxylase activity. Toxicol Appl Pharmacol 90:69–78.

Voorman R, Aust SD (1988) Inducers of cytochrome P450 d: influence on microsomal catalytic activities and differential regulation by enzyme stabilization. Arch Biochem Biophys 262:76–84.

Walker CH (1990) Persistent pollutants in fish-eating sea birds – bioaccumulation, metabolism and effects. Aquat Toxicol 17:293–324.

Watanabe I, Kashimoto T, Tatsukawa R (1986) Confirmation of the presence of the flame retardant decabromobiphenyl ether in river sediment from Osaka, Japan. Bull Environ Contam Toxicol 36:839–842.

Watanabe I, Kashimoto T, Tatsukawa R (1987) Polybrominated biphenyl ethers in marine fish, shellfish and river and marine sediments in Japan. Chemosphere 16:10–12.

Watanabe I, Tatsukawa R (1990) Anthropogenic brominated aromatics in the Japanese environment. Proceedings Workshop on Brominated Flame Retardants, Skokloster, 24–26 October 1989, KEMI, National Chemicals Directorate, Sweden, pp 63–71.

Wester PG, de Boer J (1993) Determination of polychlorinated terphenyls in biota and sediments with gas chromatography/mass spectrometry. In: Proceedings of the 13th International Symposium on Dioxins '93, Vienna. Fiedler H, Frank H, Hutzinger O, Parzefall W, Riss A, Safe S (eds) Organohalogen compounds, 14:121–124.

Wolf M, Rimkus G (1985) Chemische Untersuchungen zu einem Fischsterben mit

Hilfe der Gaschromatographie/ Massenspectrometrie. Dtsch Tierärztl Wschr 92: 174-178.

Zitko V, Hutzinger O (1976) Uptake of chloro- and bromobiphenyls, hexachloro- and hexabromobenzene by fish. Bull Environ Contam Toxicol 16:665-673.

Zitko V (1977) Uptake and excretion of chlorinated and brominated hydrocarbons by fish. Fish Mar Service Tech Rep 737, Biological Station, St. Andrews, New Brunswick.

Manuscript received March 25, 1994; accepted May 4, 1994.

Swedish Pesticide Policies 1972–93: Risk Reduction and Environmental Charges

George Ekström* and Vibeke Bernson*

Contents

I. Introduction

Recent decisions by the Swedish parliament and government to reduce the use of agricultural pesticides by 75% over 10 years have received international attention. Steps taken to reduce the risks connected with the use of nonagricultural pesticides, which constitute the majority of pesticides used in Sweden, have not attracted the same attention. The Swedish Ordinance on Pesticides defines the term "pesticide" as "a chemical product that is intended for use to protect against damage to property, sanitary nuisances, or other comparable nuisances caused by plants, animals, or microorganisms." The ordinance covers not only agricultural pesticides but also wood preservatives used by industry. Of 8914 metric tons of pesticide active ingredients (a.i.s) sold in 1993, 4856 tons (54%) consisted of creosote, a wood preservative used by industry. Industrially used chromium compounds

---

*The National Chemicals Inspectorate, P.O. Box 1384, 17127 Solna, Sweden.

© 1995 by Springer-Verlag New York, Inc.

Reviews of Environmental Contamination and Toxicology, Vol. 141.

make the second largest contribution to the overall consumption of pesticides in Sweden.

This review, therefore, has two objectives: (1) The first objective is to describe the ways and means the Swedish government and other authorities have used over two decades to reduce the risks connected with the use of pesticides not only in agriculture and horticulture but also in forestry and applications for industrial wood preservation. The central pesticide control authority [formerly the Toxic Substances Board, the Products Control Board (PCB), and, since 1986, the National Chemicals Inspectorate] has been authorized to collect detailed data on pesticide sales only since 1976, whereas Statistics Sweden has estimated purchase of pesticide products from the 1950s onward (PCB 1976b, Statistiska Centralbyrån 1990). The years selected for qualitative comparisons were 1993 and 1981, the first of five consecutive years selected by the government as a basis for comparison in the reduction programs. (2) The second objective is to describe and put into context past, present, and proposed charges on pesticides in Sweden.

Although recently regulated through a special Act and Ordinance [Swedish Code of Statutes (SCS) 1991b,c] as well as through regulations from the National Chemicals Inspectorate (KemI 1991d, KemI 1994f), biological pesticides are not discussed in this context.

II. Ban on Aerial Spraying

During the 1970s, few environmental issues attracted as much attention from both politicians and the general public as the question of whether to allow aerial spraying of pesticides and fertilizers over forests and other land generally accessible to the public by way of the traditional Right of Common Access (Sw. allemansrätten, see Lagerqvist and Lundberg 1991). The following section summarizes the legal framework of the period 1972 onward and describes available data on sprayed areas [National Board of Forestry (NBF) 1980,1993]. Of the total area of the country (45 million ha), forests account for 22.5 million ha; arable land and gardens, 2.9 million ha; permanent meadows and pastures, 600,000 ha; and remaining land and lakes, 18.5 million ha.

In a cabinet meeting in March 1972, the Swedish government empowered the Minister of Agriculture to establish a committee to investigate the use of pesticides and fertilizers as well as to study the effects and consequences of a reduced use of pesticides. On the same day, the government decided to present a bill to parliament with a proposal to ban aerial spraying of pesticides over forests and other nonagricultural land. Following the government's proposal, an act was established by parliament in April 1972, thereby, for the first time, restricting the application of pesticides from airplanes. The act made it possible for the Toxic Substances Board to grant exceptions for scientific tests and for use against pests or plant diseases (SCS 1972).

In May 1974, the 1972 Committee on the use of pesticides proposed to the government that the Act should be revoked [Ministry of Agriculture (MoA) 1974]. In June 1975, the Act was abolished by the parliament, thereby again allowing aerial spraying from July 1 the same year. In the years that followed, aerial spraying activities were controlled through a number of regulations made by the PCB (1975a,b,c), all of them based on the 1973 Act and Ordinance on products hazardous to health and the environment (SCS 1973a,b). Before spraying is allowed to be started over areas with public access, the police and health authorities as well as the public should be properly informed, the latter group through local newspapers. In addition, the land area to be sprayed should be clearly marked. In May 1976, additional provisions were added for aerial spraying over forests (PCB 1976a).

In June 1980, a new period started during which spraying over forests with pesticides intended for deciduous growth (brushwood) was not allowed (MoA 1980; SCS 1980,1981,1982a). Until December 31, 1982, the PCB could grant exceptions for scientific tests. Between January 1, 1983 and December 31, 1983, the County Administration could grant exceptions if the forest had little significance to the public's outdoor life, nature conservation, and the comfort of the local residents, and if the forest could not otherwise be cultivated according to the Act on Forestry in an economically acceptable way (SCS 1982b). From January 1984 onward, the Regional Board of Forestry could grant exemptions. However, the local authorities could decide that such exemptions may not be granted (SCS 1983). Aerial spraying, if allowed, was acceptable only with pesticides that had been specifically approved for that purpose by the PCB [Swedish Environmental Protection Agency (SEPA) 1984].

In January 1986, spreading of pesticides from aircraft was prohibited through a new ordinance (SCS 1985b). However, aerial spraying over forests with herbicides intended for brushwood growth was still covered through separate legislation already mentioned (SCS 1983,1985c). Therefore, four periods with varying degrees of restrictions may be distinguished:

(1) Between mid-1972 and mid-1975, aerial spraying was prohibited in forestry but allowed in agriculture. Data on sprayed agricultural land are unavailable for this period.

(2) Between mid-1975 and mid-1980, aerial spraying was permitted in forestry and agriculture. In forestry, the following areas were sprayed with herbicides in 1976 through 1979: 22,828; 30,078; 29,756; and 35,320 ha, respectively. The area sprayed in 1979 corresponded to 0.1% of the total forest area. In agriculture, the following arable land areas were sprayed with fungicides or insecticides during 1976–1978: 35,632; 45,715; and 46,000 (estimated) ha, respectively (Bouvin, personal communication). The area sprayed in 1977 and 1978 corresponded to 1–2% of the total arable land area.

(3) From mid-1980 until the end of 1985, aerial spraying was allowed in

agriculture but prohibited in forestry, with provisions for exemptions. No aerial spraying took place in 1980–1982. In 1983–1985, 2998, 807, and 56 ha of forests were sprayed, respectively. No information is available on aerial spraying in agriculture.

(4) From 1986 onward, aerial spraying was prohibited in agriculture and forestry. Despite provisions for exemptions in forestry, no aerial spraying took place during 1987–1991.

The ban on aerial spraying in agriculture and forestry does not seem to have resulted in severe drawbacks for these branches of society due to *inter alia* changes in attitudes, techniques, and practices.

III. Risk Reduction Programs for Agricultural Pesticides

In 1988, the Swedish parliament decided on a risk reduction program for pesticides [Swedish Government Bill (SGB) 1988]. A detailed action program aiming at a reduction of risks to health and the environment had already been prepared by the Swedish Board of Agriculture (SBA), the Environmental Protection Agency (EPA), and the National Chemicals Inspectorate (SBA 1986). The risk reduction program comprised: Changeover to pesticides with less risks; regulations on handling and use of pesticides; training in safe handling and use of pesticides; extended pesticide residue monitoring of foods and drinking water; and reduced use of pesticides.

In 1990, a new food policy was passed by the Swedish parliament (SGB 1990). The policy included provisions for further risk reduction and another 50% reduction of pesticides used in agriculture, horticulture, and forestry in 1990–1996.

The main goal of the government's and parliament's reduction programs for agricultural pesticides has been, and still is, to reduce risks connected with the use of pesticides in agriculture, horticulture, and forestry. At a very early stage, it was decided to use the number of kilograms of a.i. as a feasible means for following the progress. Therefore, the reduction aim was formulated as a 50% reduction of quantities in 5 yr (1986–1990), followed by a second 50% reduction of quantities in the next 5 yr (1991–1995). Other measures of risk reduction have been discussed by Hurst (1992). The risk reduction components and the results of the first 50% reduction program have been described by several authors in Sweden (Bernson 1991; Bernson and Ekström 1991; Emmerman 1991; Pettersson 1992, 1993, 1994; Rosengren 1991) and abroad (Anonymous 1992; Hurst 1992; Hurst et al. 1991; Weinberg 1990).

Thus far, the risk reduction programs have resulted in *inter alia* reduction in use of pesticide a.i.s in agriculture of 2900 metric tons or 67% compared with the 1981–1985 average. A very large part of the reduction can be referred to reduced use of herbicides on cereals due to the use of low-dose herbicides and reduced dose rates, reduced area of cereal cultivation, and reduced sprayed area. The SBA has estimated that lower dose

Table 1. Reduction of annual sales of pesticides by user category, in metric tons of active ingredient.

User category	1981–1985 average	1993	Reduction
Agriculture	4385	1464	67%
Horticulture	152	66	57%
Forestry	24	14	42%
Industry	8153	7019	14%
Households	832	352	58%
Total	13,546	8915	34%

Source: KemI (1994b).

rates of herbicides used against weeds in cereals correspond to approximately 1200 tons, whereas reduced cultivation of cereals and reduction of sprayed area correspond to approximately 700 tons of herbicides (Franzen, personal communication). Tables 1 and 2 show the reduction of annually sold quantities from the 1981-1985 average to 1993 by user category and type of pesticide, respectively. Quantities of pesticide a.i.s sold for agricultural and horticultural use in 1986-1993 also are shown in Figures 1 and 2, respectively.

The coordinated and wide approach taken is one important reason that the first reduction program was successful (Figure 3). Contributions have been made by SEPA (1984, 1989, 1992), the National Board of Occupational Safety and Health[3] (NBOSH) (1988), SBA (coordinating agency) (1986,1989,1991a,b,1992a,1993a,b,1994a,b), the National Food Administration (NFA) (1990, 1993a,b,c; Sandberg and Erlandsson 1990), the National Chemicals Inspectorate (KemI 1989c,1992b,1993a; Bernson 1988, 1989,1991; Bernson and Ekström 1991), and the Swedish University of

Table 2. Reduction of annual sales of agricultural pesticides by type, in metric tons of active ingredient.

Type of pesticide	1981–1985 average	1993	Reduction
Herbicides	3829	1273	67%
Fungicides	621	283	54%
Insecticides	210	52	75%
Plant growth regulators	83	41	51%
Seed dressings	161	76	53%
Total	4904	1725	65%

Source: KemI (1994b).

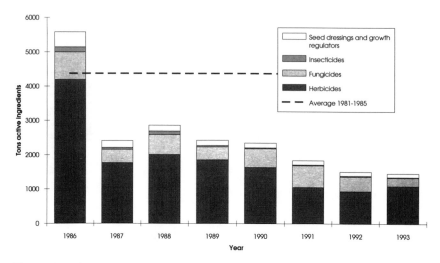

Fig. 1. Quantities of pesticides sold for use in agriculture 1986–1993, in metric tons of active ingredient.

Agricultural Sciences (Johnson et al. 1992; Kreuger and Brink 1988; Torstensson 1988; Åkerblom et al. 1990). Research at the University has been important in many areas of the reduction programs. New pesticides are tested in field trials at the University. The adjustment of pesticide doses that has occurred would have been difficult without the identification of

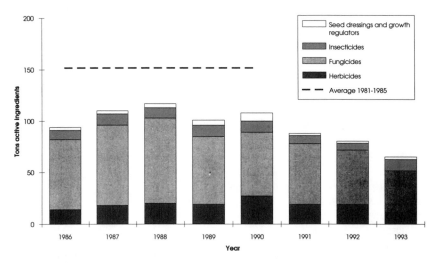

Fig. 2. Quantities of pesticides sold for use in horticulture 1986–1993, in metric tons of active ingredient.

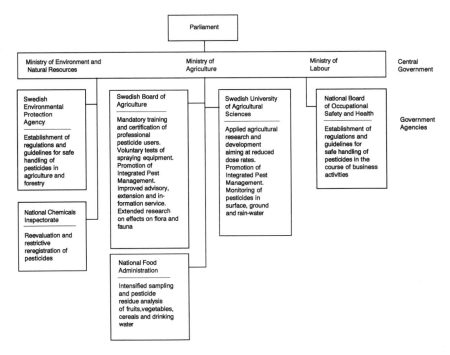

Fig. 3. Risk reduction factors in the public sector and main activities.

agro-ecological prerequisites. Techniques for pest prognoses and spraying have been improved as a result of research at the University.

An important part of the achieved reduction may be attributed to a general decrease in applications. This reduction has been particularly pronounced for herbicides. Because weed pressure in grain crops has decreased as a long-term effect of herbicide use, it is now possible to use lower doses of herbicides and still maintain weed populations at acceptable levels without creating increasing problems for the future. As pesticide doses are lowered, reliable spraying techniques become increasingly important.

A combination of economic and legal measures, applied agricultural research and technical development, and strengthened extension activities are other important elements, as was the active support by the Federation of Swedish Farmers [Svenska Lantmännens Riksforbund (LRF)]. The LRF has advertised its intent to reduce pesticide use further and to create "the world's cleanest farming" (LRF 1993; Magnusson 1993). The LRF also is promoting "environmentally friendly agriculture" on the international arena [LRF/International Federation of Agricultural Producers (IFAP) 1992; LRF 1994]. The support from the farmers and their Federation has been acknowledged publicly by the National Chemicals Inspectorate (KemI 1991b).

A second action program has been prepared jointly by the designated

Table 3. Factors affecting the future use of agricultural pesticides in Sweden and estimated maximum remaining reduction potential, in metric tons of active ingredient.

Reduction factors	Estimated reduction potential
Second 50% reduction program activities	200–300
Phaseout of remaining EBDC use on potatoes	200–300
Change in use of arable land	200
"Green tax" and other pesticide charges	70–180
Commodity prices	80
Total	750–1060

Source: SBA (1992b).

authorities (SBA 1992a). The activities of the first program will continue and, in many cases, expand. The estimated maximum remaining reduction potential is shown in Table 3. Activities with increased priority include (Franzen, personal communication): (1) Establishment of phase-out plans for unacceptable pesticides; (2) reassessment, improvement, and possible strengthening of existing regulations and general recommendations on applying pesticides; (3) improved training in safe use of pesticides; (4) applied research and improved extension service aiming at alternatives to traditional plant protection in potato cultivation; and (5) extended environmental information activities directed to officials at regional and local levels.

IV. Banned and Severely Restricted Pesticides

According to the Ordinance on Pesticides (SCS 1985b), a pesticide product when approved shall be assigned to one of the following *restriction classes*: Class 1 — pesticides that may be used only in the course of business activities by individuals holding a special permit; Class 2 — pesticides that may be used only in the course of business activities; or Class 3 — pesticides that may be used by anyone.

The number of approved pesticide products that were actually sold has decreased from 596 in 1981 to 371 in 1993 (the total number of approved products was slightly larger). The number of products in Class 3 has declined dramatically (−175). The number of products in Class 2 also has declined significantly (−68). The increase in the number of products in Class 1 (+18) reflects the fact that several products in Class 2 have been reassigned to Class 1 due to a more stringent hazard classification. The quantity of pesticides in Class 1 increased in agriculture due to the reassignment of some Class 2 products and in industry due to both a true increase in pesticide use and to the reassignment of several Class 2 products. The use of products in Class 2 decreased due to a true decrease in the use of agricultural pesticides and to the reassignment already mentioned. The use of pesticide products in Class 3 also decreased due to reduced use of pesticides in households.

Table 4 shows the number of pesticides available in approved products intended for use in agriculture, horticulture, or forestry in 1981 and 1993, respectively, by hazard category according to the World Health Organization's (WHO) classification (International Programme on Chemical Safety 1994). Few pesticides in the two most hazardous groups were available for these use categories.

Over the years, a number of pesticide chemicals have been removed from the list of approved pesticides due to the fact that they have been superseded by more effective and/or less hazardous pesticides. Therefore, they have been either voluntarily withdrawn from the market by the manufacturer or, in the latter case, been severely restricted or "banned" through administrative decisions by the competent authority. On two occasions, political deci-

Table 4. Number of pesticide active ingredients available on the market in 1981 and 1993, respectively, by WHO hazard class and use category.

WHO hazard class[a] and year		Agriculture	Horticulture	Forestry	All use categories mentioned
Ia	1981	3	4	0	6
	1993	2	3	0	5
Ib	1981	5	7	0	7
	1993	5	2	0	6
II	1981	26	19	7	32
	1993	20	18	2	29
III	1981	26	18	4	35
	1993	19	8	2	30
AHU	1981	38	43	12	59
	1993	40	24	5	49
Unclassified					
	1981	4	6	2	10
	1993	1	1	0	2
Total					
	1981	102	97	25	149
	1993	87	56	9	121

Sources: KemI (1994b); SEPA (1982).

[a]Ia = extremely hazardous; Ib = highly hazardous; II = moderately hazardous; III = slightly hazardous; AHU = unlikely to present acute hazard in normal use.

sions made by the central government have prohibited or severely restricted the use of individual pesticides. In 1977, use of the herbicide 2,4,5-T was prohibited (SCS 1977), and in 1987 the use of plant growth regulators was temporarily prohibited on cereals (SCS 1987). In 1988, this prohibition was extended indefinitely for all cereals except rye (SCS 1988a). In 1991, the exception for rye was extended to the end of 1997 (SCS 1991e).

The Chemicals Inspectorate has established regulations on pesticides that have been banned or withdrawn or for which first-time registration has been denied and therefore may not be included as a.i.s in approved pesticide products in Sweden (KemI 1993b). These regulations also contain a list of pesticides that should not be granted reregistration if or when less hazardous alternatives become available.

Sweden actively participates in a number of international activities to increase awareness of or to minimize or phase out unacceptable chemicals, including pesticides. Pesticides that have been banned, severely restricted, or voluntarily withdrawn have been identified to The United Nations Food and Agriculture Organization (FAO) and to the International Register of Potentially Hazardous Chemicals of the United Nations Environment Programme (UNEP/IRPTC) according to the London Guidelines and the Prior Informed Consent procedures (KemI 1993a; FAO/UNEP 1991; UNEP 1989). IVT, The National Association of Agrochemical and Wood Preservative Manufacturers, has established recommendations on the marketing of pesticides, based essentially on the International Code of Conduct on the Distribution and Use of Pesticides (IVT 1989; FAO 1990).

In 1989, a Swedish government official proposed a "sunset procedure" to identify pesticides and other hazardous chemicals for possible action by governments (Wahlström 1989; Foran 1991). Following that proposal, the Chemicals Inspectorate has recently presented a list of one hundred "sunset candidates" (KemI 1994a,d,e). The registration status of a number of pesticides appearing on this list and a selection of others are shown in Appendix A. The health hazard classification of the World Health Organization (WHO) has been included for comparison (IPCS 1994). In several cases, however, the environmental hazard rather than the health hazard is the reason for inclusion of substances in various lists. Neither Sweden, the Commission of the European Communities, nor UNEP have developed a similar comprehensive and administratively useful classification of pesticides according to environmental hazard (EEC 1978a,1992; KemI 1989b,1991e,1992c).

Appendix A shows that, of the pesticides currently on the FAO/UNEP PIC list (FAO/UNEP 1992), the Pesticide Action Network's "Dirty Dozen" list (PAN 1991,1992), and the European Union's "negative list" (EEC 1978b), all are either banned, voluntarily withdrawn, or unregistered in Sweden. Of the 27 pesticides listed by the National Chemicals Inspectorate as potential sunset candidates, 17 are banned or unregistered and 10 currently are in use in Sweden as insecticides, wood preservatives, or antifouling agents. Of the 36 pesticides covered by Appendices 1A and 1B(c) of the North Sea Convention (North Sea Conference 1990), 25 are banned, withdrawn, or otherwise unregistered and 11 are currently approved for use as pesticides, including wood preservatives and antifouling agents, in Sweden. Of all pesticides mentioned in Appendix A (56 entries), 20 have been banned, 4 have been voluntarily withdrawn by the manufacturer, 17 are unregistered, and 15 are approved for specified, sometimes severely restricted, uses (all other uses are prohibited) (KemI 1994c).

V. Major Pesticides Sold in 1993

Major agricultural and nonagricultural pesticides sold in Sweden in 1993 are shown in Table 5. Of the individual or groups of pesticides shown (seven entries), three are used only in industry. The four others are used in agriculture as well as for certain other purposes. Together, they accounted for 81% of the total quantity of pesticides sold in Sweden in 1993. Some of these have properties that make them hazardous to human health or the environment to such an extent that the National Chemicals Inspectorate, without having banned them, is intent to restrict further use or even to phase out their use altogether.

Creosote, a distilate of coal tar produced by high-temperature carbonization of bituminous coal, is used to impregnate wood to protect it against rot, insects, and certain marine organisms. Creosote may have serious effects on both health and environment. The use of creosote entails risks of exposure in the working environment and of diffuse dispersion, especially in soil and water [Swedish Wood Preservation Institute (SWPI) 1990]. Products containing creosote are classified as toxic and are placed in restriction class 1. All products must be labeled with risk phrases for allergy and cancer. Creosote is one of 13 compounds included in a list of chemicals that may have particularly harmful effects on the environment put together by the Chemicals Inspectorate and SEPA on the Government's request (KemI 1991a).

Chromium compounds are used to impregnate wood against rot, insects, and marine organisms. All products are classified as toxic and strongly corrosive and they are placed in restriction class 1. Chromium compounds also are classified as hazardous to the environment (KemI 1989b). Chromium(VI) compounds are allergenic and, if inhaled, carcinogenic. There is no reliable system available for collection and destruction of discarded treated wood.

Copper and copper compounds are used as antifouling agents (Debourg et al. 1993; KemI 1988), wood preservatives, and as fungicides against fungi on fruits, berries, and potatoes. They are classified as hazardous to the environment (KemI 1989b). The WHO has classified copper compounds used as agricultural pesticides as moderately and slightly hazardous (class II

Table 5. Major pesticides in 1993, in metric tons of active ingredient sold. Percentages of total quantity sold in parentheses.

Pesticide CAS Number	Function	Quantity sold, 1993	Quantity sold, 1981	Use category 1993
Creosote 8001-58-9	Wood preservative	4856 (54%)	2430[a] (24%)	Industry
Chromium compounds[b]	Wood preservative	713 (8.0%) equivalent to 350 t of chromium	546 (5.3%) equivalent to 274 t of chromium	Industry
Copper and copper compounds[c]	Wood preservative, antifouling agent, and fungicide	609 (6.8%) equivalent to 361 t of copper	613 (6.0%) equivalent to 313 t of copper	Industry, households, horticulture, and agriculture.
MCPA 94-74-6	Herbicide	311 (3.5%)	1524 (15%)	Agriculture, industry, and households
Glyphosate 1071-83-6	Herbicide	276 (3.1%)	120 (1.2%)	Agriculture, horticulture, forestry, industry, and households
Arsenic pentoxide 1303-28-2	Wood preservative	275 (3.1%)	549 (5.3%)	Industry
Ethylene bisdithiocarbamates[d]	Fungicide	171 (1.9%)	434 (4.2%)	Agriculture, horticulure, and forestry
Total	—	7211 (81%)	6216 (61%)	—

Sources: KemI (1994b); SEPA (1982).

[a]Including 1259 metric tons of tar oils.

[b]Chromium trioxide [1333-82-0] and sodium dichromate [10588-01-9] in 1981 and 1993. In 1981 also potassium dichromate [7778-50-9].

[c]Copper naphthenate [1338-02-9], copper(II)oxide [1317-38-0], copper oxychloride [1332-65-6], copper sulfate [7758-99-8], and oxine copper [1380-28-6] in 1981 and 1993. In 1981 also cupric acetate [142-71-2] and copper carbonate [1184-64-1]; in 1993 copper powder [7440-50-8], copper hydroxide [20427-59-2], copper(I)oxide [1317-39-1], copper thiocyanate [1111-67-7], and tetraminecopper [16828-95-8].

[d]Mancozeb [8018-01-7] and maneb [12427-38-2] in 1981 and 1993. In 1981 also nabam [142-59-6] and zineb [12122-67-7].

or III compounds respectively) (IPCS 1994). There is no corresponding WHO classification for nonagricultural pesticides.

MCPA is a chlorine-containing phenoxy acid herbicide classified by WHO as slightly hazardous (Class III) (IPCS 1994).

Glyphosate is a phosphonic acid herbicide classified by WHO as unlikely to present acute hazard in normal use.

Arsenic pentoxide is used to impregnate wood against rot, insects, and certain marine organisms. Studies on contamination of soil and groundwater with arsenic, copper, and chromium have been performed by the SWPI (1983,1989,1992). Arsenic pentoxide is classified as hazardous to the environment (KemI 1989b). Pesticide products are classified as toxic and strongly corrosive and are placed in restriction class 1. Products should be labeled with a risk phrase warning for carcinogenicity. No reliable system is available for collection and destruction of discarded treated wood. The government has declared that the use of arsenic and chromium compounds in wood preservatives must be reduced (SGB 1991). If the Chemical Inspectorate's regulations on preservative-treated wood (KemI 1990) do not result in a substantial reduction, the Government will propose to the parliament that an environmental tax be introduced on such wood preservatives.

The currently used ethylene bisdithiocarbamates (EBDCs) mancozeb and maneb are used as fungicides on potatoes and onions. They are classified as carcinogenic. The conversion product ethylene thiourea (ETU) is also classified as carcinogenic. The WHO has classified mancozeb and maneb as technical products irritating to skin on multiple exposure but unlikely to present acute hazard in normal use (IPCS 1994). In Sweden, products containing mancozeb or maneb are classified as toxic due to both substances' intrinsic properties and to that of ETU, and they are placed in restriction class 1. A phasing-out plan for pesticide products containing EBDCs has been established (Bergkvist 1994). EBDCs are widely used in Swedish agriculture mainly against potato blight, (*Phytophtora infestans*), sometimes at doses of 30 kg a.i./ha and growing season. The plan consists of several activities with a main goal to phase out the EBDCs entirely before the 1995 spraying season.

The Chemicals Inspectorate has declared its intent not to grant continued approval for pesticide products containing arsenic compounds, chromium compounds, mancozeb, or maneb if all present uses can be phased out, i.e., if acceptable alternatives are introduced that can replace these much needed products (KemI 1993c).

VI. Restricted Use of Preservative-Treated Wood

In September 1988, the Swedish Government commissioned NBOSH to propose actions aiming at a reduced risk connected to the use of nonagricultural pesticides, including wood preservatives. In June 1989, the Board, jointly with SEPA and the National Chemicals Inspectorate, as part of a risk reduction program proposed that the Inspectorate should establish regulations and general recommendations for the use of wood impregnated with wood preservatives (NBOSH 1989).

In December 1989, the Government commissioned the Inspectorate and the Environmental Protection Agency to make joint proposals for activities aiming at a reduced use of such chemical compounds that may have a particularly serious effect on the environment. In June 1990, the Inspectorate and the Environmental Protection Agency in their report (KemI 1991a) identified and made proposals for 13 compounds or groups of compounds hazardous to health or the environment. Arsenic, arsenic compounds, and creosote were among the chemicals identified.

The only use of creosote and the main use of arsenic in Sweden is as a wood preservative, which requires approval from the National Chemicals Inspectorate. In 1990 and 1991, respectively, the Inspectorate established regulations and general recommendations on preservative-treated wood (KemI 1990,1991c). The regulations came into force in June 1992 and aim at all those who use preservative-treated wood. The regulations are applicable to wood or other wood-based materials that have been treated with preservatives to protect against attacks by bacteria, fungi, insects, or marine organisms. The regulations contain special restrictions on the use of wood treated with wood preservatives containing arsenic, chromium, and creosote and also general requirements concerning precautionary measures in the use of all kinds of preservative-treated wood. As one consequence, preservative-treated wood may not be supplied from a treatment plant or used professionally unless the wood is dry on the surface and in all essentials not sticky, or before fixation, if any, is completed.

Wood treated with preservatives containing compounds of arsenic or chromium may be used only when long-term protection is required, such as (1) when the wood is buried in, or otherwise in permanent contact with, damp soil or water; (2) when the wood is used for the construction of jetties or other marine applications; (3) when the wood is permanently installed as a safety device to protect against accidents; and (4) when the wood is used for the interior of constructions such that it is difficult to replace and where there is a risk of accidental wetting, e.g., ground plates on plinths and concrete slabs, ground-floor joists, etc. All other use of such treated wood is prohibited. Wood treated with creosote may, for the first 30 yr after impregnation, be used professionally only and only for railway sleepers (ties) or for round timber (power poles) for transmission lines or in marine installations. Thereafter, the treated wood also may be used in certain other applications and by consumers.

In May 1993, in a letter to the Government, the Inspectorate gave an account of the effects of the regulations on preservative-treated wood. The Inspectorate concluded that because wood preservatives containing copper, chromium, and arsenic compounds are placed in the most severe restriction class with requirements for training and a permit from the NBOSH (1988), the occupational risks are under control. The environmental impact from impregnated wood used as intended is insignificant. The main problem, therefore, is related to the handling of discarded wood.

In December 1993, the Inspectorate established regulations on certain a.i.s in pesticide products (KemI 1993b). As a consequence of the regulations currently approved and recently reregistered, wood preservatives containing arsenic or chromium compounds may not be reregistered after 1998 if all current uses can be discontinued. Table 6 shows quantities of wood preservatives sold for use in industry between 1989 and 1993.

In 1992, the volume of impregnated wood was 486,000 m³, with 80% impregnated with water-soluble wood preservatives, 15% with creosote, and the remainder with fat-soluble products. Of the total production, 149,600 m³, or 31%, were exported (NBF 1993).

VII. Nonactive Ingredients

A nonactive or inert ingredient is defined as a substance contained in a preparation that by itself will not add materially to effectiveness for the purpose for which the preparation is intended. Pesticide products containing harmful inert ingredients must be labeled with information on the identity and concentration (or concentration range) of the harmful ingredient(s). Decisions on labeling requirements are made by the National Chemicals Inspectorate on a case-by-case basis during the approval proce-

Table 6. Quantities of wood preservatives sold for use in industry 1989–1993, in metric tons.

Wood preservative	1989	1990	1991	1992	1993	1993 as percentage of 1989
Arsenic pentoxide	1090	755	641	381	275	25
Chromium trioxide and sodium dichromate	1121	716	624	608	713	64
Copper(II)oxide and copper sulphate	661	441	375	381	421	64
Tetramine-copper	73	211	81	81	68	93
Creosote	4445	3588	3814	4868	4856	109
Other wood preservatives	175	66	76	255	397	227
Total	7565	5777	5611	6574	6730	89

dure. Otherwise, inerts are not publicly disclosed by the manufacturer. For adjuvants added by farmers to the spray tank to enhance pesticide performance or to improve mixing and application characteristics (Fisher et al. 1991), there are no approval requirements.

In 1992, a total of 16,271 metric tons of agricultural and nonagricultural pesticide products were sold in Sweden; 8693 tons were a.i.s and 7578 tons were inert ingredients, of which 2160 tons were volatile organic compounds. In 174 cases, decisions had been made on labeling requirements affecting 26 individually identified compounds and 12 complex mixtures of hydrocarbon solvents (of a total of 780 inert ingredients). According to a parliamentary decision, the use of volatile organic compounds in society shall be reduced by 50% between 1988 and 2000 (SGB 1991).

Over the past few years, a number of inert ingredients have been terminated due to their adverse effects on human health or the environment (acetaldehyde, benzene, chloroform, 2-ethoxyethanol, 2-ethoxyethyl acetate, 2-methoxyethanol, 2-methoxyethyl acetate, tetrachloroethene, tetrachloromethane, 1,1,1-trichloroethane, trichloroethene, vinyl chloride, and freons). A number of inert ingredients have been listed for termination, e.g., aromatic hydrocarbon solvents, dichloromethane, dimethyl formamide, formaldehyde, and isophorone (Ohlsson, personal communication).

Manufacturers and importers must observe the provisions of the Act on Chemical Products (Section 5) stating that chemical products (substances and preparations) should be avoided when less hazardous substitutes are available (SCS 1985a). Manufacturers and importers are given guidance through previous decisions made by the Parliament or Government to ban, severely restrict, or phase out chemicals (SCS 1988b,1991d; SGB 1991) and from lists of hazardous chemicals published by the Chemicals Inspectorate (KemI 1989b,1991a) or SEPA (1990).

VIII. Registration Fees and Other Pesticide Charges

Four types of pesticide charges have been in use in Sweden: an application fee, an annual registration fee (both payable to the Chemicals Inspectorate to cover its costs for approval and register-keeping of pesticides), a price regulation fee, and an environmental tax. The application fee should be paid for all pesticide products to which the Ordinance on pesticides applies (SCS 1985b; KemI 1989d). On application for approval, the applicant must pay 10,000 Swedish Kronor (SEK) for each product plus 30,000 SEK for any new a.i. not contained in an already approved product. The annual registration fee, which must be paid for each approved product, amounts to 1.8% of the sales value of the product the previous year (minimum fee 2000 SEK, maximum 200,000 SEK). The annual fee must be paid by the person who has had the pesticide product approved or by the person acting in his place.

In March 1983, a government-appointed committee on the use of pesticides and fertilizers in agriculture and forestry proposed the introduction of pesticide charges to finance a number of activities also proposed by the committee, such as improved training of pesticide users, strengthening of the central pesticide registration authority, and development of low-intensity plant protection methods, etc. (MoA 1983). After having considered risk-based, quantity-based, value-based, and area-based pesticide charges, the committee settled for a system based on quantities of pesticide a.i. sold. The committee also considered charges related to the hazard class based on health hazard only of the pesticide product, e.g., 1 SEK/kg a.i. for pesticides in the lowest hazard class, 4 SEK in the next hazard class, and 7 SEK in the highest hazard class. In the committee's opinion, the charge should burden all approved uses and all registered products, except wood preservatives. In June 1984, the Parliament instituted an act on "pesticide charges" (the environmental tax) (SCS 1984). A uniform charge of 4 SEK/ kg of a.i. was imposed on all pesticide products except wood preservatives.

In July 1986, a price regulation fee of 29 SEK/ha and applied dose was introduced for agricultural pesticides (SCS 1986). This fee was intended to reduce the use of pesticides, to cut surplus production of cereals, and to cover economic losses due to export of surplus grain and oilseeds. During 1986 and 1987, two Government-appointed working groups assessed the use of pesticides and fertilizers with an aim at reduced agricultural "intensity" and a reduced grain surplus (MoA 1986a,b,1987). The use of pesticide charges as an economic instrument was discussed by both groups. One group preferred charges differentiated according to hazard. However, considering the large variation in pesticide formulations and the existence of combinations of a.i.s of different hazards in one product, such a differential charge was considered difficult to accomplish in reality. Therefore, the working group proposed that the existing charge of 4 SEK/kg a.i. and the price regulation fee of 29 SEK/ha and application should be raised to a higher but unspecified level. The environmental tax was subsequently raised to 8 SEK in July 1988, and the price regulation fee was raised to 38 SEK in November 1990 and to 46 SEK in March 1991. In July 1991, it was lowered to 29 SEK, and finally abolished in December 1992 (SBA 1992b).

In a July 1990 report to the Government, the Environmental Charge Commission (MoE 1990) proposed that the environmental charge should be reconstructed for pesticides that were subject to both a price-regulation fee and the environmental charge. The new environmental charge should be related to the ascertained dose/ha for the pesticide concerned. The commission also proposed that this charge should be set at 40 SEK during a transitional period. The charge should cover all uses, including agriculture, forestry, horticulture, and other sectors. For pesticides that could not be included in the system (e.g., rodenticides), the commission proposed that the current charge of 8 SEK /kg a.i. be retained for the time being. The commission considered tentative charges on two levels: 40 SEK/ha and applied dose for pesticides products with relatively low risks and 500 SEK for products with high risks, and possibly, a third level at 100–150 SEK/

ha and applied dose for products with intermediate risks. A new act on environmental charges on pesticides should be enacted and made active from July 1, 1991. The commission also proposed that the National Chemicals Inspectorate should be given the task of devising detailed differentiated charges according to the hazard of different pesticides. Such a system should come into effect no later than 1995.

The Environmental Charge Commission also proposed a charge of 200 SEK/kg of arsenic and 100 SEK/kg of chromium in wood preservatives with effect from January 1, 1992. The Commission estimated that the combined effects of the proposed charges and certain other measures would reduce the use of arsenic and chromium from roughly 400 metric tons/yr to 50 tons, and from roughly 250 tons/yr to 25 tons, respectively. The government has declared that it is premature to make a decision on the introduction of additional incentives to encourage development of alternative wood preservatives (SGB 1994a).

The turnover of the Swedish market for plant protection products (agricultural pesticides) amounts to approximately 400 million SEK (equivalent to 50 million U.S. dollars) in 1991 (Resvik, personal communication). The Swedish market corresponds to approximately 0.2% of the world market in terms of value. Pesticide charges constituted about 20% of the sales value. The price regulation fee generated 80 million SEK in 1991 ($10 million U.S.). The environmental tax generated 20 million SEK ($2.5 million U.S.) in 1991 and was used to finance research and extension work (SBA 1992b). In 1993, the environmental tax generated approximately 12 million SEK ($1.5 million U.S.), equivalent to 2.5% of the sales value (Resvik, personal communication). The application fee and the annual registration fee generated 6.3 million SEK in fiscal year 1990/91, 8.7 million SEK in 1991/92, and 11 million SEK in 1992/93.

The SBA has estimated that the price regulation fee in combination with the environmental levy resulted in a reduced use of pesticides amounting to 80–180 metric tons, particularly fungicides and insecticides (SBA 1992b).

IX. Proposed Basis for a Hazard-Related "Green Tax"

The purpose of the Act on Chemical Products is to prevent injury to human health or the environment being caused by the inherent properties of pesticides and other chemical substances (SCS 1985a). A pesticide may be approved only if it is acceptable from the human health and environmental protection standpoints and is needed for the purposes stated in the Ordinance on Pesticides (SCS 1985b). However, this condition does not mean that pesticides that have been approved are harmless or without risk. The use of pesticides, therefore, can lead to various nuisances. Pesticide spraying must always be performed in such a way that humans are not harmed or caused inconvenience and so that the unintended environmental impact is minimized. Risks and nuisances may be put into one of four categories (MoA 1983): (1) occupational health risks to persons handling pesticides;

(2) risks of unintended impact on the agricultural and natural environment; (3) health risks due to unintentional ingestion of pesticide residues in food and drinking water; and (4) inconveniences due to unintended spray drift from adjacent agricultural land.

Figure 4 shows steps for assessing acceptability of agricultural pesticides in Sweden. The procedure for pesticide approval has been described by Bernson (1988,1989). The scientific and administrative components of the regulatory system have been compared internationally (Bernson 1993; GAO 1993; Schmidt-Bleek 1993). Guideline levels for clearly unwanted properties and cut-off criteria for particularly serious properties of pesticides have been published by the National Chemicals Inspectorate as a policy document (Andersson et al. 1992); see Appendices B and C. If a pesticide product contains an a.i. that is shown to have particularly serious properties according to cut-off criteria, it should not be approved. If a pesticide product contains an a.i. with clearly unwanted properties and the exposure assessment is unfavorable, the risk is considered to be unacceptable and the product should not be approved. Even if a pesticide product contains an a.i. that does not have unacceptable properties, the risk-benefit analysis may result in the conclusion that it is an unacceptable product, which should not be approved.

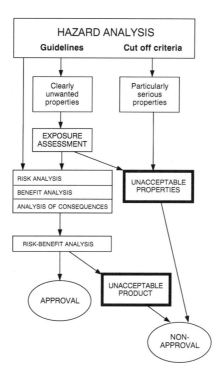

Fig. 4. Steps for assessing unacceptability of agricultural pesticides in Sweden.

To further reduce the risks connected with the use of pesticides, the Government in June 1991, commissioned the National Chemicals Inspectorate to ascertain the possibility of introducing a differential, hazard-based environmental levy on pesticides. The new levy should be based on the hazard of the product and the risk connected to the use of the product. A working group established by the Inspectorate interpreted the task as follows:

1. The principal goal of the levy is increased safety in the use of pesticides;
2. The levy should be based on unacceptable properties of pesticides (a.i.s as well as formulated products);
3. The levy should be related to the intrinsic/inherent properties or hazard and to the risks connected to the use;
4. Both environmental effects and health effects should be considered;
5. Only chemical pesticides and only pesticides used in agriculture, horticulture, forestry, parks, and households were included in the assignment.

When constructing the basis for a differential, hazard-based environmental levy, strict scientific requirements must be put on the construction (whether simple mathematical model or complex procedure) as well as on the input data and the documentation to be used. Background data should be produced by standardized methods. Data must be reliable, relevant to agricultural and climatic conditions in the country, and possible to verify. Therefore, data should be produced in accordance with OECD Guidelines and Good Laboratory Practice (SCS 1991a; OECD 1992).

Models for the assessment of a pesticide product's health and environmental hazard often comprise large numbers of parameters. However, when the model is intended for the ranking of hazards to be used as a basis for fees or taxation, only a strictly limited number of significant parameters may be used to make the process administratively simple and, consequently, relatively inexpensive. The numerical value of the selected parameters then may be used directly or converted to an index or score (O'Bryan and Ross 1988; DMU 1991; Oller et al. 1980; Reed 1985; Weber 1977; Welch 1982; Zitko 1990). A scoring system was used by the working group for ranking of both pesticides and models. However, scores were not reproduced in the final report.

Based on the government's instructions and its own interpretations, the working group constructed three models based on selections of a limited number of parameters related to health and environmental effects. In addition, a fourth "procedure" consisting of an expert evaluation was studied and compared with the simple models. In the first model, the Inspectorate's existing hazard classes for health effects of pesticide products were combined with parameters for environmental effects based on an existing system for testing and hazard evaluation in the aquatic environment, "ESTHER" (Gabring 1988; Landner 1987; KemI 1989a). In the second model, the working group selected two simple parameters related to the environmental effects and combined them with the Inspectorate's hazard classes for health effects. In the third model, the ESTHER Manual's parameters for environmental effects were combined with the health hazard classification recommended by the WHO (IPCS 1994) and the acceptable daily intake (ADI) values recommended by the Joint FAO/WHO Meetings on Pesticide Residues (WHO 1992). The fourth "procedure" consisted of a tentative assessment of health and environmental effects based on existing Swedish criteria for identifying clearly unwanted and particularly serious properties of pesticides (Andersson et al. 1992; Johnson, undated) (see also Appendices B and C).

After having studied the outcome for a selection of pesticides and compared the results of the three models and the fourth procedure, the working group found that the third model, based on certain internationally produced input data, was particularly unsuitable. ADI values and hazard classifications are not updated frequently enough to keep up with the constant flow of toxicity data made available to national registration authorities. Therefore, this model was excluded from further considerations. In a second phase, models 1 and 2 and the "procedure" were tested on a wider selection of pesticides. The working group concluded that model 1, based on the ESTHER Manual and existing hazard classification for pesticide products, was relatively fast but also relatively coarse due to the fact that the Manual was originally intended for screening purposes and does not include several parameters relevant to the assessment of pesticides. Model 2, based on potential bioaccumulation (K_{ow}), aquatic toxicity, and the existing health hazard classification, was considered to be administratively simple but crude due to the exclusion of several important factors, such as mobility, persistence, and toxicity to bees. The expert evaluation, based on the Inspectorate's criteria for health and environmental effects, was considered to result in the best outcome despite the fact that specific criteria for allergic and environmental effects, such as toxicity for fish, bees, and birds, do not exist. However, this procedure, which includes a hazard assessment of intrinsic properties and a risk assessment based on exposure, is very time consuming and therefore expensive.

The working group in its report (KemI 1992b) proposed, as a first-hand alternative, that pesticides should be grouped at three levels according to their potential health and environmental effects. Pesticides with properties that exceed one or more cut-off criteria should be placed, by the National Chemicals Inspectorate, at the highest hazard level. Pesticides that have properties that exceed two or more guideline levels should be placed at an intermediate level. Other pesticides should, in principle, be put on the lowest hazard level. As a result of this differentiation, levies also could be put at three levels (Table 7). As a second-hand alternative, the working group proposed a levy based on standard dose rates only. The construction of such a levy would be identical to the former price regulation fee.

The working group's report was handed over to the government as an

Table 7. Main principles of the proposed hazard-based environmental levy.

Hazard levels and criteria[a]	Registration decision	Proposed undifferentiated levy aiming at general reduction of quantities used	Proposed differentiated levy aiming at substitution with less hazardous pesticide or nonchemical method
One or more cut-off criteria exceeded	Nonapproval; Phaseout of approved product	0–10 SEK/dose[b]	150–200 SEK/dose
Two or more guideline levels exceeded. Exposure assessment	Nonapproval or approval[c]	0–10 SEK/dose	50 SEK/dose
Maximum one guideline level exceeded	Approval	0–10 SEK/dose	No levy

[a]Guideline levels and cutoff criteria are shown in Appendices B and C, respectively.

[b]One "dose" is the amount of a pesticide active ingredient recommended by the manufacturer to be applied on one hectare.

[c]Decision based on the availability of alternative pesticides and alternative pest control methods, and the agricultural demand for the pesticide.

Appendix to a report from the SBA (1992b), who had been given the main responsibility for an assessment of the effects of the environmental levy and the price regulation fee on the use of pesticides and fertilizers. In the latter report, the SBA proposed a basic charge of maximum 10 SEK/dose and higher steering charges of 50 and 150–200 SEK, respectively, on the intermediate and highest hazard levels. Awaiting the necessary classification to be made by the Chemicals Inspectorate, a uniform charge of 20 SEK/dose was proposed by the Board. No decision on hazard-related charges has been made by the Government as of November 1994. However, the Government has declared that it will decide on confining environmental levies for and environmental classification of pesticides when the first reregistration phase has been concluded (SGB 1994a). In a recent bill to the parliament, the Government has proposed a provisional increase of the environmental charge from 8 SEK/kg a.i. to 20 SEK/kg a.i. pending a new system (SGB 1994b).

Summary

The prime goal of the Swedish pesticide policies has been, and still is, to reduce the risks connected with the use of pesticides. However, the actual means used for achieving these goals differ between agricultural and nonag-

Table 8. Summary of differences and similarities in Swedish pesticide policies for agricultural and certain nonagricultural pesticides (wood preservatives), April 1994.

Parameter	Agricultural pesticides	Wood preservatives
Approval and registration by	The National Chemicals Inspectorate	The National Chemicals Inspectorate
Pesticides mainly produced	Abroad	Abroad
Directions for use issued by	The National Chemicals Inspectorate, the National Board of Occupational Safety and Health and the Swedish Environmental Protection Agency	The National Chemicals Inspectorate and the National Board of Occupational Safety and Health
Training and permits	Swedish Board of Agriculture (pesticide products in class 1L and 2L)	National Board of Occupational Safety and Health (pesticide products in class 1 Ass)
Guideline levels and cut-off criteria established for toxic effects and environmental fate	Yes	No
Export of pesticide-treated products	Insignificant	Extensive
Quality standards and/or use restrictions for pesticide-treated products	Maximum residue limits established for foods by the National Food Administration	Use restrictions for preservative-treated wood established by the National Chemicals Inspectorate
Covered by first and second 50% reduction programs 1988 and 1990	Yes	No
Covered by other risk reduction program(s) 1989	No	Yes (see KemI 1991a; NBOSH 1989)
Current environmental tax	Yes (8 SEK per kilogram of active ingredient)	No
Quantity-based tax proposed or considered by commission	Yes	Yes
Differential, hazard-based "green tax" proposed by working group	Yes	No (not in assignment)

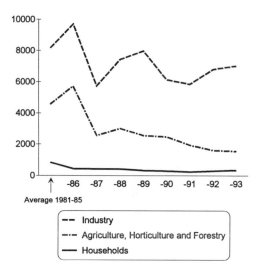

Fig. 5. Quantities of pesticides sold by use category 1986–1993, in metric tons of active ingredient.

ricultural pesticides (Table 8). Important elements of the agricultural pesticide policy are the two 50% reduction programs between 1986 and 1990, and between 1991 and 1996, respectively. Measured as kg of a.i.s, the reduction in use of agricultural pesticides so far has reached 67% of the scheduled 75% (Figure 5).

However, hazardous pesticides are still in use both in industry (creosote, arsenic, and chromium compounds) and in agriculture (e.g., EBDCs). Therefore, risk reduction programs should continue, aiming at a use reduction of the most hazardous pesticides still in use. A differential and hazard-based "green tax" for chemical pesticides of any type or intended use may be a future incentive for substitution of pesticides and possibly also for further reduction in the use of agricultural and nonagricultural pesticides.

Acknowledgment

The assistance of Magnus Franzen, Ann-Marie Ekström, Maud Erkhammar, and Per-Håkan Wistrand in producing the figures is gratefully acknowledged.

References

Åkerblom M, Thoren L, Staffas A (1990) Determination of pesticides in drinking water [in Swedish with English summary], Vår Föda 42(4–5):236–243.

Andersson L, Gabring S, Hammar J, Melsäter B (1992) Principles for Identifying Unacceptable Pesticides, The National Chemicals Inspectorate, Report No. 4, Solna.

Anonymous (1992) Swedish risk reduction plan extended, Agrow World Crop Protection News, 173:10–11.

Bergkvist P (1994) Phasing out plan for maneb and mancozeb in Sweden [in Swedish with English summary], Swedish University of Agricultural Sciences, Proc 35th Swedish Crop Protection Conference, pp 77–80.

Bernson V (1988) Regulation of pesticides in Sweden. In: Proc Brighton Crop Protection Conference, Pests and Diseases Vol 3, British Crop Protection Council, pp. 1059–1064.

Bernson V (1989) Experiences and reflections on registration and reregistration of pesticides, In: Seminar on Pesticides in Agriculture, EC-EFTA Meeting organized by the Swedish Ministry of Environment, Uppsala 14–15 November, The National Chemicals Inspectorate, Solna, pp. 81–86.

Bernson V (1991) The Swedish fifty percent cut-off, In: Meeting on the Control of Pesticides Proc, Saltsjöbaden (Sweden), October 29–31, The National Chemicals Inspectorate, Solna, pp. 113–117.

Bernson V, Ekström G (1991) Swedish policy to reduce pesticide use. Pesticide Outlook 2(3):33–36.

Bernson V (1993) The role of science in pesticide management — An international comparison: The Swedish experience. Reg Toxicol Pharmacol 17:249–261.

Debourg C, Johnson A, Lye C, Törnqvist L, Unger C (1993) Antifouling products (A summary of health and environmental effects), Report No 2/93, The National Chemicals Inspectorate, Solna.

Danmarks Miljøundersøgelser (1991) Environmental impact of pesticides: Perspectives for use of an environmental index [in Danish], Report from a group of experts, Danmarks Miljøundersøgelser, Soeborg, January 1991.

European Economic Communities (EEC) (1967) Council Directive on the approximation of laws, regulations, and administrative provisions relating to the classification, packaging, and labeling of dangerous substances (67/548/EEC), European Economic Communities, Brussels, June 27.

EEC (1978a) Council Directive on the approximation of the laws of the Member States relating to the classification, packaging and labeling of dangerous preparations (pesticides) (78/631/EEC), European Economic Communities, Brussels, June 26.

EEC (1978b) Council Directive prohibiting the placing on the market and use of plant protection products containing certain active substances (79/117/EEC), European Economic Communities, Brussels, December 21.

EEC (1992) Council Directive amending for the seventh time Directive 67/548/EEC on the approximation of the laws, regulations, and administrative provisions relating to the classification, packaging and labeling of dangerous substances (92/32/EEC), European Economic Communities, Brussels, April 30.

Emmerman A (1991) Programme to reduce the risks of pesticides in Sweden. Pesticides News 14:12–14, December.

Food and Agriculture Organization (FAO) (1990) International Code of Conduct on the Distribution and Use of Pesticides (Amended version), Food and Agriculture Organization of the United Nations, Rome.

FAO/United Nations Environment Programme (UNEP) (1991) Guidance for Governments, Operation of the Prior Informed Consent Procedure for Banned or Severely Restricted Chemicals in International Trade, Joint FAO/UNEP Programme for the Operation of Prior Informed Consent, Food and Agriculture Organization of the United Nations, and the United Nations Environment Programme, Rome and Geneva.

FAO/UNEP (1992) Fourth FAO/UNEP joint meeting on Prior Informed Consent (PIC), Geneva February 17-21, (Report) United Nations Environment Programme and Food and Agriculture Organization of the United Nations, Geneva.

Fisher N, Vlahovski F, Weil ED, Rigo WA, Jr (1991) Farm Chemicals Handbook '91, Meister Publishing Company, Willoughby, OH.

Foran JA (1991) The Sunset Chemicals Proposal, International Environmental Affairs, pp. 303-308.

Gabring S (1988) Initial Assessment of the Environmental Hazard of Chemical Substances: An Evaluation of the "ESTHER Manual" [in Swedish with English summary], KemI Report No. 9/88, The National Chemicals Inspectorate, Solna.

General Accounting Office (GAO) (1993) A Comparative Study of Industrialized Nations' Regulatory Systems, Report to the Chairman, Committee on Agriculture, Nutrition and Forestry, US Senate, US General Accounting Office, Ref No GAO/PEMD-93-17, Washington, DC.

Hurst P, Hay A, Dudley N (1991) Pesticide Handbook, Journeyman Press, London.

Hurst P (1992) Pesticide Reduction Programmes in Denmark, the Netherlands, and Sweden, WWF International Research Report, World Wide Fund For Nature, Gland, Switzerland.

International Programme on Chemical Safety (1994) The WHO Recommended Classification of Pesticides by Hazard, and Guidelines to Classification 1994–1996, WHO/PCS/94.2, The World Health Organization, Geneva.

Industrin för Växt-och Träskydd (IVT) (1989) Recommendations on the marketing of pesticides [in Swedish], Industrin för Växt-och Träskyddsmedel, Stockholm.

Johnson A (undated) Environmental risk management in the Swedish reregistration programme for pesticides (Leaflet), The National Chemicals Inspectorate, Solna.

Johnson A, Kreuger J, Lundbergh I (1992) Pesticides and Surface Water – A Review of Pesticide Residues in Surface Waters in the Nordic Countries and Problems Related to Pesticide Contamination, paper presented at the 2nd meeting of the Council of Europe ad hoc group of experts on pesticides and the environment, Paris April 28-29, The National Chemicals Inspectorate, Solna.

KemI (1986) The National Chemical Inspectorate's regulations on classification and labeling, KIFS 1986:3.

KemI (1988) The National Chemical Inspectorate's regulations with respect to antifouling products, KIFS 1988:3.

KemI (1989a) Systems for Testing and Hazard Evaluation of Chemicals in the Aquatic Environment, A Manual for an Initial Assessment – "ESTHER", Report No 4/89, The National Chemicals Inspectorate, Solna.

KemI (1989b) Chemical compounds hazardous to the environment, A collection of examples with scientific documentation [in Swedish with English summary], Report No 10/89, The National Chemicals Inspectorate, Solna.

KemI (1989c) The National Chemicals Inspectorate's regulations on proficiency requirements for use of certain pesticides [in Swedish], KIFS 1989:6.

KemI (1989d) The National Chemical Inspectorate's regulations on application of the Ordinance (1985:836) on pesticides, KIFS 1989:7.

KemI (1990) The National Chemicals Inspectorate's regulations on preservative-treated wood, KIFS 1990:10.

KemI (1991a) Risk Reduction of Chemicals, A government commission report from the National Chemicals Inspectorate and the Swedish Environmental Protection Agency, KemI Report No 1/91. The National Chemicals Inspectorate, Solna.

KemI (1991b) You made it possible! Acknowledgment to the Swedish farmers who reduced the use of pesticides with the support from the Board of Agriculture, the Chemicals Inspectorate and the Environmental Protection Agency [in Swedish], The National Chemicals Inspectorate, Solna.

KemI (1991c) The National Chemical Inspectorate's general recommendations (1991:4) to the regulations on preservative-treated wood.

KemI (1991d) The National Chemical Inspectorate's regulations on the sale and use of biological pesticides, KIFS 1991:7.

KemI (1991e) Seminar on Environmental Classification and Labeling of Chemicals, EC-EFTA Meeting, Uppsala, March 20–21, The National Chemicals Inspectorate, Solna.

KemI (1992a) Act on Chemical Products, English translations of the Act and connected Ordinances. The National Chemicals Inspectorate, Solna.

KemI (1992b) Hazard-based environmental levies on pesticides [in Swedish], Report from a working group on risk assessment, December. The Swedish National Chemicals Inspectorate, Solna.

KemI (1992c) The National Chemicals Inspectorate's regulations on classification and labelling in connection with transfer of chemical substances dangerous for the environment, KIFS 1992:2.

KemI (1993a) Pesticides notified to FAO and IRPTC as having been banned, severely restricted, or voluntarily withdrawn in Sweden. The National Chemicals Inspectorate, September 3.

KemI (1993b) The National Chemicals Inspectorate's regulations on certain pesticides (a.i.s) [in Swedish], KIFS 1993:5.

KemI (1993c) Labeling of chemical products, Regulations and general advice with regard to the classification and labeling of chemical products hazardous to health, flammables and explosives [in Swedish], The National Chemicals Inspectorate and the National Inspectorate of Explosives and Flammables, Solna.

KemI (1994a) Selecting multiproblem chemicals for risk reduction, A presentation of the Sunset project, The National Chemicals Inspectorate, Solna.

KemI (1994b) Quantities of pesticides sold 1993 [in Swedish with a guide in English], The National Chemicals Inspectorate, Solna.

KemI (1994c) The National Chemicals Inspectorate's list of approved pesticides 1994 [in Swedish with a guide in English].

KemI (1994d) Chemical Substances Lists: A guide to the lists used in the Swedish Sunset Project, Report No 10/94, The National Chemicals Inspectorate, Solna.

KemI (1994e) Selecting multiproblem chemicals for risk reduction: A presentation of the Swedish Sunset Project, Report No 13/94, The National Chemicals Inspectorate, Solna.

KemI (1994f) The National Chemicals Inspectorate's regulations on biological pesticides, KIFS 1994:4.

Kreuger J, Brink N (1988) Losses of pesticides from arable land, Växtskyddsrapporter — Jordbruk 49, 50–61, Swedish University of Agricultural Sciences, Uppsala.

Lagerqvist M, Lundberg E (1991) The Right of Public Access: An asset at risk? [in Swedish with English abstract], Report TRITA-KUT 91:3051, The Royal Institute of Technology, Stockholm.

Landner L (1987) Environmental Hazard of Chemicals, Manual for an initial assessment — "ESTHER" [in Swedish], SNV Report 3243, Swedish Environmental Protection Agency, Solna.

LRF (Federation of Swedish Farmers) (1993) We are creating the world's cleanest farming, Federation of Swedish Farmers advertisement in: Tomorrow — Global Environment Business No 3, July–September, p 21.

LRF (1994) Farmers of the World: Close to the Problem, Close to the Solution, Federation of Swedish Farmers advertisement in: Tomorrow–Global Environment Business No 2, April–June, pp 54–57.

LRF/IFAP (1992) First Session of the Environment Committee, International Federation of Agricultural Producers, Proc Stockholm December 3–4, Federation of Swedish Farmers.

Magnusson G (1993) The World's Cleanest Farming: The Swedish Farmers' Vision [in Swedish], Miljö och Hälsa 4:20–21.

Ministry of Agriculture (MoA) (1974) Use of pesticides and fertilizers [in Swedish], Report by the Committee on the use of pesticides and fertilizers, Swedish Official Report Series 1974:35, Ministry of Agriculture, Stockholm.

MoA (1980) Control of brushwood [in Swedish with English summary], Report by the Committee on the use of pesticides and fertilizers in agriculture and forestry, Ds Jo 1980:11, Ministry of Agriculture, Stockholm.

MoA (1983) Control of Pests and Weeds [in Swedish], Report by the Committee on the use of pesticides and fertilisers in agriculture and forestry, Swedish Official Report Series 1983:11, Ministry of Agriculture, Stockholm.

MoA (1986a) Short-term measures to reduce the grain surplus [in Swedish], Report by a group of experts on cereal cultivation, Ds Jo 1986:2, Ministry of Agriculture, Stockholm.

MoA (1986b) Measures to reduce the grain surplus [in Swedish], Report by a group of experts on cereal cultivation, Ds Jo 1986:6, Ministry of Agriculture, Stockholm.

MoA (1987) Agricultural productivity, environmental impact and grain surplus [in Swedish], Report by the Working group on low-intensity agricultural production, Ds Jo 1987:3, Ministry of Agriculture, Stockholm.

MoE (Ministry of Environment) (1990) Environmental charges and other economic instruments [in Swedish with English summary], Report by the Environmental Charge Commission, Swedish Official Report Series 1990:59, Ministry of Environment, Stockholm.

National Board of Forestry (NBF) (1980) Statistical Yearbook of Forestry [in Swedish], Official Statistics of Sweden, National Board of Forestry, Jönköping.

NBF (1993) Statistical Yearbook of Forestry [in Swedish with English summary], Official Statistics of Sweden, National Board of Forestry, Jönköping.

National Board of Health and Welfare (NBHW) (1983) Regulations on permits for use of certain pesticides [in Swedish], SoSFS 1983:16, National Board of Health and Welfare, Stockholm.

National Board of Occupational Safety and Health (NBOSH) (1988) Regulations and general recommendations on safe use of pesticides [in Swedish], AFS 1988: 5, The National Board of Occupational Safety and Health, Solna.

NBOSH (1989) Reduced risks from the use of wood preservatives [in Swedish], Joint National Board of Occupational Safety and Health/ National Chemicals Inspectorate/ Environmental Protection Agency Report, Report No 1989:4, National Board of Occupational Safety and Health, Solna.

National Food Administration (NFA) (1990) Pesticide residues in food [in Swedish], Report by a working group on food quality, SLV Report No 1990:14, The National Food Administration, Uppsala.

NFA (1993a) Pesticide residues in food of plant origin 1992, SLV Report No 12, The National Food Administration, Uppsala.

NFA (1993b) The National Food Administration's regulations and recommendations on pesticide residues in food [in Swedish], SLV FS 1993:32.

NFA (1993c) The National Food Administration's regulations and recommendations on drinking water [in Swedish], SLV FS 1993:35.

North Sea Conference (1990) Final Declaration of the Third International Conference on the Protection of the North Sea, Final Declaration, The Hague, March 8.

O'Bryan TR, Ross RH (1988) Chemical scoring system for hazard and exposure identification, J Toxicol Environ Hlth 24(1):119–134.

Organization for Economic Cooperation and Development (OECD) (1992) The OECD Principles of Good Laboratory Practice, OECD Series on Principles of Good Laboratory Practice and Compliance Monitoring No 1, Environment Monograph No 45, Organisation for Economic Cooperation and Development, Paris.

Oller WL, Cairns T, Bowman MC, Fishbein L (1980) A toxicological risk assessment procedure: A proposal for a surveillance index for hazardous chemicals, Arch Environ Contam Toxicol 9:483–490.

Pesticide Action Network (PAN) (1991) Demise of the Dirty Dozen: Chart, background information and source material, Pesticide Action Network, North America Regional Center, San Francisco.

PAN (1992) Demise of the Dirty Dozen, Global Pesticide Campaigner, May, p 12.

Products Control Board (PCB) (1975a) The Products Control Board's regulations on information requirements when using pesticides in areas with public access [in Swedish], PKFS 1975:8.

PCB (1975b) The Products Control Board's regulations on aerial spraying of pesticides [in Swedish], PKFS 1975:9.

PCB (1975c) The Products Control Board's regulations on the use of herbicides in forestry [in Swedish], PKFS 1975:10.

PCB (1976a) The Products Control Board's regulations on the use of pesticides in areas with public access [in Swedish], PKFS 1976:1.

PCB (1976b) The Products Control Board's regulations on the obligation to report quantities of pesticides sold [in Swedish], PKFS 1976:3.

Pettersson O (1992) Pesticides, valuations, and politics, (Discussion note) J Agric Environ Ethics 103–106.

Pettersson O (1993) Swedish pesticide policy in a changing environment, In: Pimentel D, Lehman H (eds), The Pesticide Question, Chapman & Hall, New York, pp 182–205.

Pettersson O (1994) Reduced pesticide use in Scandinavian agriculture, Crit Rev in Plant Sci, 13(1):43–55..

Reed DV (1985) The FDA surveillance index for pesticides: Establishing food monitoring priorities based on potential health risk, J Assoc Offic Anal Chem 68(1): 122–124.

Rosengren H (1991) The Swedish fifty percent cut-off. In: Meeting on the Control of Pesticides, (Proc), Saltsjöbaden (Sweden), October 29–31, The National Chemicals Inspectorate, Solna, pp 119–124.

Sandberg E, Erlandsson B (1990) Pesticide residues in surface and ground waters [in Swedish with English summary], Vår Föda 42(4–5):224–234.

Swedish Board of Agriculture (SBA) (1986) Joint Swedish Board of Agriculture/ Swedish Environmental Protection Agency/National Chemicals Inspectorate programme to reduce the risks to health and the environment from the use of pesticides [in Swedish], Swedish Board of Agriculture, Jönköping (September).

SBA (1989) Reduced chemical pest control [in Swedish], Joint Swedish Board of Agriculture/Swedish Environmental Protection Agency/National Chemicals Inspectorate/National Board of Occupational Safety and Health report, Jönköping and Stockholm, May.

SBA (1991a) Problem areas in chemical pest control: Suggested actions [in Swedish], Report No 1, Swedish Board of Agriculture, Jönköping.

SBA (1991b) Results of the action programme to reduce the risks to health and the environment from the use of pesticides [in Swedish with English summary], Joint Swedish Board of Agriculture/Swedish Environmental Protection Agency/ National Chemicals Inspectorate report, Report No 8, Swedish Board of Agriculture, Jönköping.

SBA (1992a) Health and environmental risks associated with the use of pesticides — Result of the action plan [in Swedish], Report No 34, Swedish Board of Agriculture, Jönköping.

SBA (1992b) Environmental levies on pesticides and fertilizers [in Swedish], Report No 41, Swedish Board of Agriculture, Jönköping.

SBA (1993a) Regulations on proficiency tests and permits for use of certain pesticides [in Swedish], SJVFS 1993:59, Swedish Board of Agriculture, Jönköping.

SBA (1993b) Reduced risks to human health and the environment from the use of pesticides — Result of the action plan [in Swedish], Report No 24, Swedish Board of Agriculture, Jönköping.

SBA (1994a) Why did the use of pesticides go down? [in Swedish], Swedish Board of Agriculture, Jönköping, February 3.

SBA (1994b) Programme to reduce risks connected with the use of pesticides in Sweden, Swedish Board of Agriculture, Jönköping.

Statistiska Centralbyrån (SCB) (1990) The Natural Environment in Figures [in Swedish with Swedish-English vocabulary], Third Ed., Statistics Sweden, Stockholm.

SCB (1993a) Pesticides in Swedish agriculture: Number of doses in 1992 [in Swedish with English summary], Statistiska meddelanden nr Na 31 SM 9302, Statistics Sweden, Örebro.

SCB (1993b) Use of pesticides in agriculture in 1991/1992 [in Swedish with English summary], Statistiska meddelanden nr Na 31 SM 9303, Statistics Sweden, Örebro.

Schmidt-Bleek F, Marchal M M (1993) Comparing Regulatory Regimes for Pesticide Control in 22 Countries: Toward a New Generation of Pesticide Regulations, Regul Toxicol Pharmacol 17:262–281.

Swedish Code of Statutes (SCS) (1972) Act on prohibition on certain aerial spraying of pesticides [in Swedish], Swedish Code of Statutes 1972:123.

SCS (1973a) Act on products hazardous to health or the environment [in Swedish], Swedish Code of Statutes 1973:329.

SCS (1973b) Ordinance on products hazardous to health or the environment [in Swedish], Swedish Code of Statutes 1973:334.

SCS (1977) Ordinance amending the Ordinance (1973:334) on products hazardous to health or the environment [in Swedish], Swedish Code of Statutes 1977:246.

SCS (1979) Ordinance on mercury-dressing of planting seeds [in Swedish], Swedish Code of Statutes 1979:349.

SCS (1980) Act on temporary prohibition on use of certain pesticides in forestry [in Swedish], Swedish Code of Statutes 1980:368.

SCS (1981) Ordinance on continued temporary prohibition on use of certain pesticides in forestry [in Swedish], Swedish Code of Statutes 1981:351.

SCS (1982a) Act on prohibition on certain use of herbicides in forestry [in Swedish], Swedish Code of Statutes 1982:240.

SCS (1982b) Act on the use of certain herbicides in forestry [in Swedish], Swedish Code of Statutes 1982:242.

SCS (1983) Act on the use of certain herbicides in forestry [in Swedish], Swedish Code of Statutes 1983:428.

SCS (1984) Act on pesticide charges [in Swedish], Swedish Code of Statutes 1984:410.

SCS (1985a) Act on chemical products [in Swedish], Swedish Code of Statutes 1985:426 (available in English, see KemI 1992a).

SCS (1985b) Ordinance on pesticides [in Swedish], Swedish Code of Statutes 1985:836 (available in English, see KemI 1992a).

SCS (1985c) Ordinance on the spreading of pesticides over forestland [in Swedish], Swedish Code of Statutes 1985:842 (available in English, see KemI 1992a).

SCS (1986) Act amending the act (1967:340) on price regulation in agriculture [in Swedish], Swedish Code of Statutes 1986:389.

SCS (1987) Ordinance amending the Ordinance (1985:836) on pesticides, Swedish Code of Statutes 1987:119.

SCS (1988a) Ordinance amending the Ordinance (1985:836) on pesticides, Swedish Code of Statutes 1988:77.

SCS (1988b) Ordinance on CFCs and Halons etc. [in Swedish], Swedish Code of Statutes 1988:716 (available in English, see KemI 1992a).

SCS (1991a) Ordinance on the implementation of OECD's Guidelines on Good Laboratory Practice [in Swedish], Swedish Code of Statutes, SFS 1991:93.

SCS (1991b) Act on preliminary examination of biological pesticides [in Swedish], Swedish Code of Statutes 1991:639 (available in English, National Chemicals Inspectorate).

SCS (1991c) Ordinance on preliminary examination of biological pesticides [in Swedish], Swedish Code of Statutes 1991:1288 (available in English, National Chemicals Inspectorate).

SCS (1991d) Ordinance on Certain Chlorinated Solvents [in Swedish], Swedish Code of Statutes 1991:1289 (available in English, see KemI 1992a).

SCS (1991e) Ordinance amending the Ordinance (1985:836) on pesticides, Swedish Code of Statutes 1991:1291.

Swedish Environmental Protection Agency (SEPA) (1982) Use of Pesticides 1981 [in Swedish], SNV PM 1569, Swedish Environmental Protection Agency, Solna.

SEPA (1984) The Swedish Environmental Protection Agency's regulations on the outdoor use of pesticides [in Swedish], SNFS 1984:2 (PK 19).

SEPA (1989) Use of Pesticides in Agriculture and Horticulture, General guidelines 88:2, Swedish Environmental Protecion Agency, Solna.

SEPA (1990) Strategy for volatile organic compounds (VOC) – Discharge, effects, actions [in Swedish], Report No 3763, Swedish Environmental Protection Agency, Solna.

SEPA (1992) Agriculture, Sections 11.01 B, 11.02 B, 11.03 C, and 11.04 C of the Environment Protection Ordinance (1989:364), Swedish Environmental Protection Agency, Solna, January.

SEPA (1994) Sweden free from ozone-depleting chemicals [in Swedish], Report No 4278, Swedish Environmental Protection Agency, Solna.

Swedish Government Bill (SGB) (1988) Swedish Government Bill (1987/88:128) on agricultural activities to improve the environment [in Swedish], Stockholm.

SGB (1990) A New Food Policy [in Swedish], Swedish Government Bill 1989/90: 146, Stockholm.

SGB (1991) A Living Environment [in Swedish with separate English summary of main proposals], Swedish Government Bill No 1990/91:90, Stockholm.

SGB (1994a) Policy for a continued adaptation to a recycling society: Measures to reduce chemical risks [in Swedish], Swedish Government Bill No 1993/94:163, Stockholm.

SGB (1994b) Some food policy measures associated to a membership of the European Union [in Swedish], Swedish Government Bill No 1994/95:75, Stockholm.

Swedish Wood Preservation Institute (SWPI) (1983) Contamination of soil and groundwater at wood preserving plants [in Swedish with English summary], Meddelande nr 146, Swedish Wood Preservation Institute, Stockholm.

SWPI (1989) Studies on the fixation of arsenic in soil and on the mobility of arsenic, copper, and chromium in CCA-contaminated soil [in Swedish with English summary], Meddelande nr 161, Swedish Wood Preservation Institute, Stockholm.

SWPI (1990) Sanitation of creosote-contaminated soil [in Swedish], Meddelande nr 162, Swedish Wood Preservation Institute, Stockholm.

SWPI (1992) Leakage of arsenic, copper and chromium from preserved wooden chips deposited in soil – Eleven years of field experiments [in Swedish with English summary], Meddelande nr 166, Swedish Wood Preservation Institute, Stockholm.

Torstensson L (1988) Pesticides in the environment – Appearance, transport and effects: Literature review and suggestions for research [in Swedish with English summary], SNV Report No 3536, Swedish Environmental Protection Agency, Solna.

United Nations Environment Program (UNEP) (1989) London Guidelines for the Exchange of Information on Chemicals in International Trade (Amended version), United Nations Environment Programme, Nairobi.

Wahlström B (1989) Sunsetting for dangerous chemicals, Nature 341:276.

Weber JB (1977) The pesticide scorecard. Environ Sci Technol 11(8):756–761.

Weinberg AC (1990) Reducing agricultural pesticide use in Sweden, J Soil Water Conserv 45(6):610–613.

Welch JL, Ross RH (1982) An approach to scoring of toxic chemicals for environmental effects, Environ Toxicol Chem 1:95–102.

World Health Organization (WHO) (1992) Summary of Toxicological Evaluations Performed by the Joint FAO/WHO Meetings on Pesticide Residues, WHO/PCS/92.9, The World Health Organization, Geneva.

Zitko V (1990) Priority ranking of chemicals for risk assessment, Sci Total Environ 92:29–39.

Manuscript received April 28, 1994; accepted May 10, 1994.

Appendix A. Registration Status in Sweden for Pesticides on Selected International "Blacklists."

Pesticide CAS number	WHO hazard class[a]	PIC list	Dirty Dozen list	List of potential sunset candidates	North Sea lists[b]	EEC negative list	Registration status in Sweden
Aldicarb 116-06-3	Ia		X				Banned
Aldrin 309-00-2	Ib	X	X	X	X	X	Banned
Arsenic compounds	—			X	X		Arsenic pentoxide approved for industrial wood preservation (restriction class 1)
Atrazine 1912-24-9	AHU				X		Banned
Azinphos-ethyl 2642-71-9	Ib				X		Unregistered
Azinphos-methyl 86-50-0	Ib				X		Approved for use in horticulture (restriction class 1)
Binapacryl 485-31-4	II					X	Unregistered
Camphechlor (Toxaphene) 8001-35-2	II	X	X	X	"not used"	X	Unregistered
Captafol 2425-06-1	Ia				"not used"	X	Banned

Carbofuran 1563-66-2	Ib	X				Unregistered
Carbon tetrachloride 56-23-5	(not classified)	X	X			Unregistered
Chlordane 5103-74-2	II			X	X	Banned
Chlordimeform 6164-98-3	II		"not used"	X		Unregistered
Chloropicrin 76-06-2	(not classified)		X			Unregistered
Chlorpyrifos 2921-88-2	II	X				Approved for industrial use (restriction class 1)
Chromium compounds	–	X	X			Chromic trioxide and sodium dichromate approved for industrial wood preservation (restriction class 1)
Copper and copper compounds	–	X	X			Copper hydroxide and copper oxychloride approved for use as fungicides, copper (II) hydroxide carbonate, copper naphthenate, copper(II)oxide, copper sulfate, oxine-copper, and tetramincopper as wood preservatives, and copper powder, copper(I)oxide and copper thiocyanate as antifouling agents

Appendix A. (*continued*)

Pesticide CAS number	WHO hazard class[a]	PIC list	Dirty Dozen list	List of potential sunset candidates	North Sea lists[b]	EEC negative list	Registration status in Sweden
Creosote 8001-58-9	(not classified)			X			Approved for industrial wood preservation (restriction class 1)
Cyhexatin 13121-70-5	III	X					Banned
DBCP 96-12-8	Ia	"No longer produced"	X				Voluntarily withdrawn
DDT 50-29-3	II	X	X	X	X	X	Banned
Diazinon 333-41-5	II			X			Approved for use in horticulture (restriction class 2)
1,2-Dibromo ethane (ethylene dibromide, EDB) 106-93-4	(not classified)	X	X	X	X	X	Unregistered
1,2-Dichloro ethane 107-06-2	(not classified)				X	X	Unregistered
Dichlorvos 62-73-7	Ib				X		Unregistered
Dicofol 115-32-2	III					X	Voluntarily withdrawn

Substance (CAS)	Classification					Status
Dieldrin 60-57-1	Ib	X	X	X	X	Banned
Dimethoate 60-51-5	II		X	X		Approved for use in agriculture and horticulture (restriction class 2)
Dinoseb 2813-95-8	Ib	X	X	X	X	Banned
Endosulfan 115-29-7	II		X	X		Approved for use in horticulture (restriction class 2)
Endrin 72-20-8	Ib	X	X	X	X	Banned
Ethylene oxide 75-21-8	(not classified)		X	X	X	Banned
Fenitrothion 122-14-5	II			X		Approved for use in horticulture (restriction class 2)
Fenthion 55-38-9	Ib			X		Approved for use in households (restriction class 3)
Fluoroacetic acid 144-49-0 and its derivatives	Fluoro acetamide: Ib	Fluoro acetamide		X	X	Unregistered
Heptachlor 76-44-8	II	X	X	X	X	Unregistered
Hexachlorobenzene 118-74-1	Ia	X	X	X	X	Voluntarily withdrawn
Hexachlorocyclohexane, HCH 608-73-1	II	X	X	X	X	Unregistered

Appendix A. (continued)

Pesticide CAS number	WHO hazard class[a]	PIC list	Dirty Dozen list	List of potential sunset candidates	North Sea lists[b]	EEC negative list	Registration status in Sweden
Lindane 58-89-9	II		X				Banned
Malathion 121-75-5	III				X		Approved for use in agriculture and households (restriction class 2 and 3)
Maleic hydrazide 123-33-1 10071-13-3	AHU					X	Unregistered
Mercury compounds	–	X		X	X	X	Banned or unregistered (SCS 1979)
Methyl bromide 74-83-9	(not classified)			X			Approved for industrial use (restriction class 1). To be phased out (SEPA 1994)
Nitrofen 1836-75-5	AHU				X	X	Banned
Paraquat 4685-14-7	II	X	X				Banned
Parathion, ethyl 56-38-2	Ia	X	X		X		Banned
Parathion, methyl 298-00-0	Ia	X	X	X	X		Unregistered
Pentachlorophenol 608-93-5	Ib	X	X	X	X		Banned

Compound	Classification[a]				Status / Comments
Polychlorinated terpenes 8001-50-1	(not classified)			X	Unregistered
Quintozene 82-68-8	AHU		X	X	Voluntarily withdrawn
Simazine 122-34-9	AHU			X	Approved for use in horticulture and forestry (restriction class 1 and 2)
2,4,5-T 93-76-5	II	X[c]		"not used"	Banned
Thallium and thallium compounds	Thallium sulfate: Ib		X		Banned or unregistered
Tributyltin compounds	bis(Tributyltin) oxide: Ib	bis(Tributyltin) oxide	X	X	Tributyltin methacrylate copolymer and bis(tributyltin)oxide are approved as antifouling agents, and bis(tributyltin)oxide and tributyltin naphthenate as wood preservatives
Trifluralin 1582-09-8	AHU			X	Banned
Triphenyltin compounds	Fentin acetate: II			X	Unregistered

Sources: EEC (1978b); FAO/UNEP (1992); IPCS (1994); KemI (1994a, 1994c); North Sea Conference (1990); PAN (1991).

[a] For classifications, see footnote to Table 4.

[b] "Not used," pesticide that would have been included in Annex 1B(c) but was considered not to be in current use.

[c] Applicable to 2,4,5-T with dioxin contamination.

Appendix B. Guideline Levels and Cutoff Criteria for Toxic Effects Applicable to Agricultural Pesticides.

Parameter	Guideline level	Cutoff criterion[a]
Acute toxicity	Pesticide products that, on account of their high acute toxicity, are assigned to the category Toxic have clearly unwanted properties. Pesticides with such properties are unacceptable unless it can be conclusively shown that the potential exposure is low.	Pesticide products that, on account of their very high acute toxicity, are assigned to the category Very toxic, are unacceptable.
Eye and skin irritation	Pesticide products that, on account of their corrosive properties, are assigned to the category Corrosive have clearly unwanted properties. Pesticides with such properties are unacceptable unless it can be conclusively shown that the potential exposure is low.	Pesticide products that, on account of their corrosive properties, are assigned to the category Highly corrosive are unacceptable.
Sensitization by skin contact	Pesticide products containing >0.1% of a strongly sensitizing substance have clearly unwanted properties. If a product contains <0.1% of the substance, an evaluation of the risk of sensitization must be made in the individual case. Pesticides with clearly unwanted properties are unacceptable unless it can be conclusively shown that the potential exposure is low.	None
Sensitization by inhalation	Pesticide products containing >1% of substances that may cause sensitization of the respiratory tract and that can give rise to symptoms of a not negligible type have clearly	None

unwanted properties. If a product contains <1%, a case-by-case assessment should be made. Pesticides with clearly unwanted properties are unacceptable unless it can be conclusively shown that the potential exposure is low.

Subacute toxicity (14 d study or 28 d study)	Pesticide products containing >5% of a substance that causes serious damage at the following levels have clearly unwanted properties: oral 15 mg/kg and d; dermal 30 mg/kg and d; inhalation 0.1 mg/L, 6 hr/d. Pesticides with clearly unwanted properties are unacceptable unless it can be conclusively shown that the potential exposure is low.	Pesticide products containing >5% of a substance that causes serious damage at the following levels are unacceptable: oral 3 mg/kg and d; dermal 5 mg/kg and d; inhalation 0.02 mg/L, 6 hr/d.
Subchronic toxicity (90 d study)	Pesticide products containing >5% of a substance that causes serious damage at the following levels have clearly unwanted properties: oral 5 mg/kg and d; dermal 10 mg/kg and d; inhalation 0.05 mg/L, 6 hr/d. Pesticide products with clearly unwanted properties are unacceptable unless it can be conclusively shown that the potential exposure is low.	Pesticide products containing >5% of a substance that causes serious damage at the following levels are unacceptable: oral 1 mg/kg and d; dermal 2 mg/kg and d; inhalation 0.01 mg/L, 6 hr/d.
Chronic toxicity (at least a 12 mon study)	Pesticide products containing >5% of a substance that causes serious damage at the following levels have clearly unwanted properties: oral 0.5 mg/kg and d. Pesticides with clearly unwanted properties are unacceptable unless it can be conclusively shown that the potential exposure is low.	Pesticide products containing >5% of a substance that causes serious damage at the following levels are unacceptable: oral 0.1 mg/kg and d.

Appendix B. *(continued)*.

Parameter	Guideline level	Cutoff criterion[a]
Carcinogenicity	Pesticide products that, on account of their carcinogenic properties, are assigned to the category Harmful, have clearly unwanted properties. Pesticides with such properties are unacceptable unless it can be conclusively shown that the potential exposure is low.	Pesticide products that, on account of their carcinogenic properties, are assigned to the category Toxic are unacceptable.
Mutagenicity[b]	Pesticide products containing >1% of a substance regarded as a category 3 substance have clearly unwanted properties. Within the interval 0.1–1% risks should be assessed on a case-by-case basis. Pesticides with clearly unwanted properties are unacceptable unless it can be conclusively shown that the potential exposure is low.	Pesticide products containing >1% of a substance regarded as a category 1 or category 2 substance are unacceptable. Within the interval 0.1–1.0%, a case-by-case assessment should be performed.
Reproductive toxicity	Pesticide products containing >0.5% of a substance that is toxic to human reproduction have clearly unwanted properties. Pesticide products containing >5% of a substance that is classified as toxic to reproduction on the basis of animal studies have clearly unwanted properties if the substance is toxic at doses lower than 200 mg/kg body weight and d. Within the interval 0.5–5% a case-by-case assessment should be performed. Potency and all other relevant factors should be considered.	Pesticide products containing >5% of a substance that is toxic to human reproduction are unacceptable. Pesticide products containing >5% of a substance that is classified as toxic to reproduction on the basis of animal studies are unacceptable if the substance is toxic at doses lower than 50 mg/kg body weight and d. Within the interval 0.5–5% a case-by-case assessment should be performed.

Pesticide products with clearly unwanted properties are unacceptable unless it can be conclusively shown that the potential exposure is low.

Source: Andersson et al. (1992).

[a]Toxicity categories according to KemI (1986).

[b]Criteria for classification of mutagenicity are based on EEC (1967).

Appendix C. Guideline Levels and Cutoff Criteria for Environmental Fate Parameters Applicable to Agricultural Pesticides.

Parameter	Guideline level	Cutoff criterion
Degradability	Pesticides (a.i.) or toxic transformation products thereof with degradability exceeding the following limits have clearly unwanted properties: Half-life more than 7 wk at 25 °C; 10 wk at 20 °C; 14 wk at 15 °C; or 20 wk at 10 °C. Pesticides with such properties are unacceptable if they pose, at potential levels of exposure, a not negligible risk of accumulation in the treated soil or in other parts of the environment.	A pesticide product is unacceptable if the degradability of its a.i.s or toxic transformation products thereof exceeds the following limits: 18 wk at 25 °C; 26 wk at 20 °C; 37 wk at 37 °C; or 52 wk at 10 °C.
Bioaccumulation	Pesticides (a.i.) or toxic transformation products thereof with bioconcentration factors (BCF) exceeding 500 in fish have clearly unwanted properties. Pesticides with such properties are unacceptable if they, at potential levels of exposure, pose a not negligible risk for accumulation in biota.	Pesticide products are unacceptable when their a.i.s or toxic transformation products thereof have a bioconcentration factor (BCF) exceeding 2000 in fish and a half-life in soil or water exceeding 1 mon at 20 °C.
Mobility	Pesticides (a.i.) or toxic transformation products thereof with high mobility ($K_{oc} < 150$) have clearly unwanted properties. Pesticides with such properties are unacceptable if they, at potential levels of exposure, pose a not negligible risk of reaching areas outside the application site in concentrations that may cause adverse effects in the environment.	Pesticide products are unacceptable when their a.i.s or toxic transformation products thereof are expected to have very high mobility in soil ($K_{oc} \leq 50$) and a half-life in soil exceeding 1 mon at 20 °C.

Source: Andersson et al. (1992).

Transport of Organic Environmental Contaminants to Animal Products

George F. Fries*

Contents

I. Introduction

The use of synthetic organic chemicals, particularly pesticides, has been a normal agricultural practice for many years. In addition to compounds intentionally used in agricultural production, other synthetics with the potential to cause adverse environmental effects and/or food contamination have been introduced into agricultural environments. These unintended contaminants have arisen as by-products of the synthesis of approved compounds, as contaminants of materials used in normal practice, from accidental misuse of industrial compounds in place of approved compounds,

*Meat Science Research Laboratory, Beltsville Agricultural Research Center, 10300 Baltimore Avenue, Beltsville, MD 20705-2350 U.S.A.

© 1995 by Springer-Verlag New York, Inc.

Reviews of Environmental Contamination and Toxicology, Vol. 141.

and as constituents of emissions and discharges from industrial processes. The number of potential contaminants and contamination scenarios is large. For example, any semivolatile organic compound hypothetically can be transported as a vapor or particle through the atmosphere and deposited onto plants, soils, and other surfaces (Bidleman 1988). Nonvolatile compounds may also be transported through the atmosphere in the particle phase. Hazardous organic chemicals potentially present in such land-applied materials as sewage sludge and waste water are estimated to number in the thousands (O'Connor et al. 1991; Jacobs et al. 1987). Because of the diversity of chemicals and circumstances, knowledge of the fate and transport of the unintended contaminants contains many gaps and areas of uncertainty.

Several representative classes of environmental contaminants are evaluated in this review for their potential to be transported through agricultural systems to foods of animal origin. The classes of compounds were selected to represent a range of chemical properties, which are important determinants of fate and transport in the environment. Estimation of the quantities of a contaminant transported from a point of introduction into the environment to animal products requires the evaluation of multifactor pathways involving soils, plants, and animals. Numerous pathways of animal product contamination have been identified and evaluated in the context of specific chemicals, agricultural and industrial practices, and exposure scenarios (Connett and Webster 1987; Dean and Suess 1985; Fries 1982, 1987; Fries and Paustenbach 1990; Stevens and Gerbec 1988). The important pathways include: (1) Introduction of the chemical onto plants by deposition of vapors, particles, and other matrices, and consumption of the plants by animals; (2) introduction of the chemical to soil, transfer to plants by root uptake and translocation or by volatilization from soil and deposition, and consumption of the plants by animals; (3) introduction of the chemical to soil, and ingestion of the soil by animals; and (4) introduction of the chemical into feed during processing. Pathways involving drinking water contamination and inhalation exposure are not significant in most circumstances. In addition to the properties of a chemical, agricultural management practices, such as the types of feed and the use of pastures, are important parameters of the major exposure pathways (Fries 1991; Fries and Paustenbach 1990).

II. Compounds and Sources
A. Characterization of Contamination

Situations involving contamination by synthetic organic chemicals can be placed into two categories based on the extent and duration of the potential exposures. The first category is the localized contamination that results from industrial accidents, or the application of contaminant-containing materials to limited areas. Typical examples include the release of tetrachlorodibenzo-*p*-dioxin (TCDD) in an industrial accident in Seveso, Italy (Pocchiari et al. 1983), the accidental substitution of a polybrominated biphenyl (PBB) fire retardant for a normal feed ingredient in Michigan (Fries 1985b), and the application of sewage sludge to agricultural land (Pahren 1980; Riordan 1983). The second category is the widespread contamination caused by aerial dispersion and transport of chemicals from industrial emissions and by the presence of contaminants in widely used products. Typical examples include the formation of polycyclic aromatic hydrocarbons (PAHs) during combustion (Edwards 1983), the formation of polychlorinated dibenzo-*p*-dioxins (PCDDs) and dibenzofurans (PCDFs) in incinerators (Olie et al. 1977; Bumb et al. 1980), and the presence of TCDD in phenoxy herbicides that were widely used in the past (Helling et al. 1973).

As illustrated, a given compound or class of compounds can be involved in either type of situation. The major distinction between the categories is the greater ease of characterizing, managing, and correcting localized contamination compared with widespread contamination. Localized contamination generally is more circumscribed in time and area than widespread contamination, and fewer exposure pathways may be involved. In localized situations, concentrations of the contaminants that result from accidents can be high, with corresponding high risks to a small number of individuals. In contrast, widespread contamination will result in low-level exposures in large populations.

B. Polyhalogenated Aromatic Compounds

Several classes of polyhalogenated aromatic compounds are among the most prominent environmental pollutants. Significant classes include the polyhalogenated biphenyls, PCDDs, and PCDFs, and residual chlorinated insecticides that have persisted from extensive use in the past. These compounds are lipophilic and, depending on the number and positions of halogen substitution, resistant to chemical and biological degradation. As a result, many of these compounds concentrate in biological systems and are transported to lipid-containing animal products. Public concern about these compounds is high because of their prevalence and potential toxicological effects, which include carcinogenicity and endocrine-disrupting developmental effects (Colborn et al. 1993; Davis et al. 1993; Gallo et al. 1991; Kimbrough et al. 1984).

The PCDDs and PCDFs are two classes of related compounds with similar chemical and toxicological characteristics. The number of possible positional congeners is 75 for PCDDs and 135 for PCDFs (Figure 1). Of the 210 possible congeners, only the 7 PCDDs and 10 PCDFs with chlorine substitutions in the 2, 3, 7, and 8 positions are considered to produce the characteristic "dioxin-like" toxicity (United States Environmental Protection Agency [U.S. EPA] 1989). Toxicity equivalency factors (TEF) that relate the toxicity of all congeners to TCDD, the most toxic congener, have

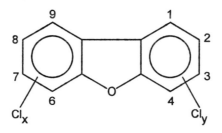

Polychlorinated Dibenzo-*p*-dioxins

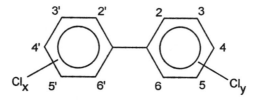

Polychlorinated Dibenzofurans

Polychlorinated Biphenyls

Fig. 1. Structures of the polychlorinated dibenzo-*p*-dioxins (PCDDs), polychlorinated dibenzofurans (PCDFs), and polychlorinated biphenyls (PCBs).

been assigned to the active congeners. Although TEFs are convenient for expressing the hazards of a mixture in a target tissue, the practice of reporting environmental data as a single toxicity equivalent quantity can be misleading because of the differing fates of the individual congeners (McLachlan 1993).

The focus on sources of PCDDs and PCDFs in the environment has changed over time. The original concerns related to the presence of these compounds as contaminants of phenoxy herbicides and other chlorophenol-based chemicals (Langer et al. 1973; Helling et al. 1973). In addition to the widespread introductions from pesticide use, PCDDs and PCDFs have also been involved in high-level localized contamination from industrial accidents and improper waste disposal (Carter et al. 1975; Pocchiari et al. 1983). It has since been determined that PCDDs and PCDFs are formed in many combustion processes when suitable precursors and conditions are present (Bumb et al. 1980; Olie et al. 1977). Municipal waste incinerators are now considered to be the major sources of PCDDs and PCDFs in aerial emissions (Fiedler and Hutzinger 1992; Harrad and Jones 1992; Rappe 1992). Hazardous waste and hospital incinerators, paper mills, metal refineries, and automobile exhausts also are considered significant sources.

The identification of PCDDs and PCDFs as products of combustion has led to the suggestion that such nonanthropogenic processes as natural forest fires may be a source of the background levels of these compounds, but evidence for this suggestion is not strong. Traces of PCDDs and PCDFs were detected in the smoke of controlled burns, but the compounds could have arisen from resuspension of background material (Tashiro et al. 1990). Analyses of ancient tissues indicate that concentrations were much lower than present background concentrations (Ligon et al. 1989). Dioxins were present in low concentrations in sediment cores as early as 1860, but the concentrations increased rapidly after 1920 and reached a peak around 1980, which suggests that industrial activity has been the major source (Czuczwa et al. 1984; Smith et al. 1992).

Aerial transport of combustion emissions has been the primary source of PCDDs and PCDFs in agricultural environments since the phaseout of phenoxy herbicides, but several minor sources may be important in localized situations. Land application has been a method for disposal of sewage and paper mill sludges, which may at times contain PCDDs and PCDFs (Bilawckuk et al. 1989; Rappe 1989; U.S. EPA 1990; Weerasinghe et al. 1985). Pentachlorophenol was a widely used wood preservative that contained PCDD and PCDF contaminants (Firestone et al. 1972). Former wood treatment sites and facilities constructed with treated wood can be potential residue sources.

The polychlorinated biphenyls (PCBs) also are a group of chemically related compounds, of which there are 209 possible congeners (Figure 1). Approximately 25 of these account for 50% to 75% of the environmental burden of PCBs (McFarland and Clarke 1989). Eleven of these, referred to as coplanar PCBs because their rings can rotate into the same plane, are believed to have dioxin-like toxicity, but these congeners are only a small portion of environmental residues. PCBs had wide applications in the electrical industry and as hydraulic and heat exchange fluids (Nisbet and Sarofim 1972). Their manufacture was phased out in the 1970s, but use continued until existing equipment became obsolete. PCBs have vapor pressures in the range of 10^{-4} to 10^{-11} atm at ambient temperatures (Bidleman 1988). Volatilization and aerial transport of the persistent congeners has led to the presence of residues throughout the environment, and low background concentrations can be found in any animal product. Concentrations of PCBs in archived herbage samples have declined significantly since 1965, coinciding with restrictions on manufacture and use since about 1970 (Jones et al. 1992). The decline in higher-molecular-weight compounds was less than that of the more labile lower-molecular-weight compounds, which suggests that present concentrations of PCBs in air are mainly due to volatilization from secondary sources such as soil and landfills.

Specific agricultural applications of PCBs were few, but significant animal residue problems have occurred because of leaking heat exchangers and transformers, use as an ingredient in a silo sealant, and misuse of waste oil (Nisbet and Sarofim 1972; Robens and Anthony 1980; Willett and Hess 1975). Sewage sludges, particularly those of industrial cities, have contained PCBs, a factor that must be evaluated when applying sludges to agricultural land (Mumma et al. 1983, 1984; Weerasinghe et al. 1985). The frequency and severity of these localized introductions have declined with the phase-out of PCBs.

A number of organochlorine insecticides were used heavily in the past, but this use has been largely discontinued since the 1970s. Many of these are persistent, and significant concentrations remain in soils where usage was high (Hitch and Day 1992). Residual organochlorines may result in significant animal residues under some management conditions (Willett et al. 1993). The organochlorine insecticides are transported globally by the same mechanisms as the dioxins and PCBs (Bidleman 1988). Thus, technical hexachlorocyclohexane (HCH) and DDT, which are used heavily in the tropics, may be transported to the U.S. and other areas where their use has been banned.

Many of the chlorinated contaminants have brominated analogs with similar chemical and toxicological characteristics. The lesser environmental significance of brominated contaminants is mainly due to the lesser industrial importance of bromine, which is used primarily in fire retardant applications. Pyrolysis of thermoplastics has led to the formation of brominated dibenzofurans and dibenzo-p-dioxins, and municipal incinerator fly ash was found to contain bromochlorodienzo-p-dioxins and dibenzofurans in amounts approximately 1–20% of the amounts of corresponding compounds in the chloro series (Donnelly et al. 1990). The PBBs, which are similar to the PCBs in structure and chemistry (Figure 1), had limited use as flame retardants in the early 1970s. PBB became a contaminant of localized significance when it was accidently mixed into cattle feed and distributed to a large number of farms in Michigan in 1974 (Fries 1985a). A significant portion of the PBB in feed passed through the animals, leading to soil contamination where manure was spread and subsequent recycling to animals (Fries and Jacobs 1986).

C. Polycyclic Aromatic Hydrocarbons

The polycyclic aromatic hydrocarbons (PAHs) are a widely distributed class of environmental contaminants, many of which are known to be mutagenic and carcinogenic (International Agency for Research on Cancer 1983; Edwards 1983). The PAHs, which are formed during incomplete combustion of organic matter, are composed of three or more benzene rings and contain only carbon and hydrogen (Figure 2). Although PAHs can arise through such natural processes as forest and prairie fires, analyses of sediment cores have suggested that the quantities entering the environment have increased since the onset of the industrial age in the 1800s, which coincides with the

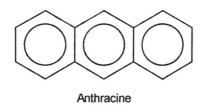

Anthracine

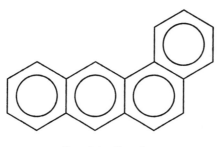

Benz(a)anthracine

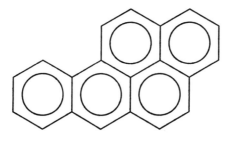

Benzo(a)pyrene

Fig. 2. Structures of representative polycyclic aromatic hydrocarbons (PAHs).

increased use of fossil fuels (Laflamme and Hites 1978; Wickstrom and Tolonen 1987; Wakeham et al. 1980). Similar conclusions were drawn from analyses of archived soil samples from an untreated plot in a rural area at the Rothamsted Experimental Station, England, that had been sampled continuously since 1846 (Jones et al. 1989c). The findings also clearly demonstrated that the primary source of PAH entry into agricultural environments is by aerial transport from combustion sources.

Concentrations of PAHs in agricultural soils near industrial urban areas can be several orders of magnitude greater than concentrations in remote rural areas, but the qualitative composition of the mixture is relatively constant (Jones et al. 1989b). Removal of PAHs by runoff and drainage water results in high concentrations in sewage sludge, with concentrations of individual PAHs in the range of 1–10 mg/kg (Wild et al. 1990a). Application of these sludges to agricultural land has produced areas of elevated soil concentrations of PAHs (Wild et al. 1990b).

D. Other Contaminants

The phthalate esters (Figure 3) are commonly used plasticizers that have been detected in a variety of environmental media, including air, water, and soils, in areas remote from industrial activity (Graham 1973; Thomas and

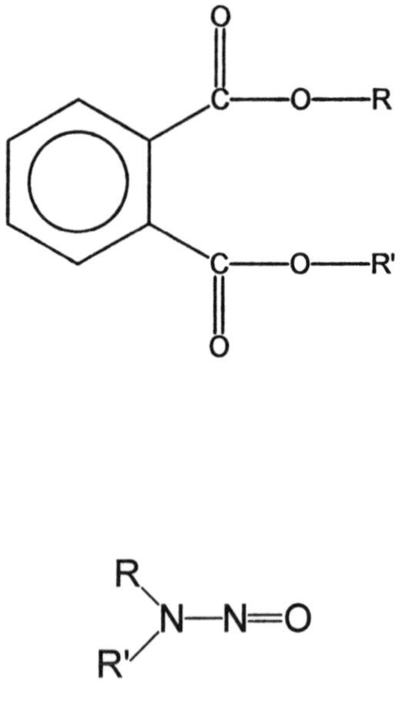

Nitrosamines

Fig. 3. Basic structures of phthalate esters and nitrosamines.

Northrup 1982). The predominant phthalate ester in use and in the environment is di(2-ethylhexyl)phthalate (DEHP). This ester also predominates in the extent of available information, but there are sufficient similarities in the available information on other esters to allow inferences to be drawn (Kluwe 1982). It is assumed that most phthalate esters in the environment arise from their volatilization during incineration or by leaching from plastics in landfills or other environmental situations (Thomas and Northrup 1982). Phthalate esters are the most prominent toxic organic in sewage sludge, and disposal on agricultural land is a potential source of localized phthalate ester contamination (Jacobs et al. 1987). The median concentrations of DEHP found in surveys of sludges were approximately 100 mg/kg dry weight.

The N-nitrosamines (Figure 3) are a group of carcinogens formed by the interaction of nitrite with secondary and tertiary amines (Mirvish 1970). They have been detected at trace concentrations in sewage sludges (Mumma et al. 1983, 1984) and in nitroanaline-based herbicides (Kearney et al 1980a). However, since amines are abundant in nature and nitrite is formed by the microbial reduction of nitrate, natural ecosystems are a major source of nitrosamines (Tate and Alexander 1976). Typically, the major concerns involving anthropogenic sources have emphasized production and processing practices that have enhanced the level of precursors (Gough et al. 1977; Kofoed 1981).

Pentachlorophenol was used extensively as a wood preservative, herbicide, and fungicide (Cirelli 1978). Although many of its uses were discontinued, treated wood was used extensively in livestock facilities (Shull et al. 1981) which remain a reservoir source of pentachlorophenol in animal production systems. Low concentrations of pentachlorophenol frequently are found in animal tissues due to exposure to these sources and from inappropriate use of treated wood shavings as bedding (Butler and Frank 1991; Ryan et al. 1985).

III. Fate and Transport in Plants and Soils
A. Aerial Transport and Deposition

Plants and soils are the points of entry of xenobiotic chemicals into agricultural environments, except in a few situations involving localized contamination. Although the physical and chemical properties of compounds are important determinants of fate and transport, the manner of introduction and the characteristics of any carriers have important effects on their fate and transport. Adsorption of compounds to various organic and inorganic matrices may have significant effects on volatility, bioavailability, and retention on plants, which must be included in the quantitative evaluation of the transport of a compound.

Aerial transport of compounds from sources and subsequent deposition is the major pathway for introduction for many environmental contaminants. The contaminants of greatest concern are semivolatile, with vapor pressures between 10^{-4} and 10^{-11} atm at ambient temperatures (Bidleman 1988). Vapor phase transport is expected to predominate in compounds with higher vapor pressures and particle transport in compounds with lower vapor pressures. This expectation was confirmed in a study of PCDDs and PCDFs in the ambient atmosphere in which the ratios of vapor-to-particle bound phases were in the range 0.01 to 1 and 30 to 1, with the ratios dependent upon vapor pressure of the compound and ambient temperature (Eitzer and Hites 1989a). These data represent compounds with vapor pressures in the range of 10^{-8} to 10^{-12} atm.

Deposition processes may be categorized as wet and dry, and each process has different implications for plant contamination and retention. Precipitation scavenging was the most efficient for particle-bound PCDDs and PCDFs (Eitzer and Hites 1989b). Particle-bound compounds have higher molecular weights and lower aqueous solubilities than those in the vapor phase because of their higher degrees of chlorination. This leads to the common observation of higher relative levels of octachlorodioxins and furans in sediments than in ambient air. Additional factors leading to enhanced concentrations include the greater susceptibility of the less chlorinated compounds in the vapor phase to photodegradation, and the Henry's law constants would predict greater vaporization of the less chlorinated congeners from the water column. The fraction of dioxins and furans observed in particle-bound wet deposition was inversely related to ambient temperature because the distribution between vapor phase and particles is directly related to ambient temperature, and the washout ratio is greater for particles (Koester and Hites 1992). The distribution between wet and dry deposition of PCDDs and PCDFs was approximately equal in two locations in Indiana (Koester and Hites 1992). However, it is expected that distributions between wet and dry deposition would differ in areas with different intensities and frequencies of precipitation. The dry deposition velocity assumed by Koester and Hites (1992) was 0.2 cm/sec, which is much lower than the 1.0 cm/sec used in some dioxin assessments (Olie et al. 1983; Connett and Webster 1987). Although this brief discussion involved only PCDDs and PCDFs, the principles should also apply to other compounds with comparable vapor pressures and Henry's law constants.

B. Vapor Uptake by Plants

Airborne chemicals in the vapor phase are accumulated on plant surfaces in cuticle waxes, which are present in all terrestrial plants. These waxes can be removed by washing with organic solvents, and nonionic organic compounds with high octanol-water partition coefficients generally are removed quantitatively with the waxes (Reischl et al. 1989). This property of plants has been used to monitor spatial and temporal variations in ambient concentrations of chlorinated hydrocarbons and PAHs (Gaggi and Bacci 1985;

Herrmann and Baumgartner 1987; Strachan et al. 1994). Needles of pine or other conifers have been used most commonly for monitoring. Concentrations of various PAHs usually are highly correlated with each other, and total concentrations are related to proximity of combustion sources. Concentrations of PCBs usually are greatest in urban and industrial areas, whereas concentrations of organochlorine insecticides are greatest in areas of past or present agricultural use.

Recently, effort has been devoted to developing predictive relationships between air concentrations and plant accumulations. Typically, this relationship is expressed as the bioconcentration factor (BCF), which is the equilibrium partition coefficient between concentrations of a compound in plant biomass and air. A good relationship between log BCF and log K_{ow}, (octanol–water partition coefficient) was demonstrated for a limited number of organochlorines (Travis and Hattemer-Fry 1988), but this finding was probably fortuitous and a model incorporating K_{aw} (air-water partition coefficient), a dimensionless Henry's Law constant (Suntio et al. 1988), has been proposed as more appropriate (Bacci et al. 1990a,b). The BCFs of a number of pesticides and environmental contaminants on leaves of azalea plants maintained in a chamber with constant concentrations in air were determined and the following relationship was found.

$$\log (BCF\ K_{aw}) = -1.95 + 1.14 \log K_{ow} \quad R = 0.92. \tag{1}$$

This relationship indicates that highly lipophilic compounds require low K_{aw} values (i.e., high vapor pressures) in order to demonstrate significant bioaccumulation, whereas polar organic compounds may have high BCFs even if their vapor pressure is low. Further refinements of the model led to the conclusion that the octanol–air partition coefficient (K_{oa}) is the key partitioning descriptor for plant accumulation of lipophilic compounds from air (Paterson et al. 1991). Experimental K_{oa} values have not been measured, but it was suggested that this parameter could be estimated from K_{ow}/K_{aw}.

Although the azalea leaf model appears appropriate for demonstrating general principles, it is likely that the constants derived in Eq. (1) are species-specific and may not be useful for predicting accumulation on plants of agronomic significance. For example, the concentrations of PCB accumulated on the leaves of 18 plant species collected at the same site varied greatly, with a range as much as two orders of magnitude (Buckley 1982). The uptake of vapor phase TCDD by reed canarygrass (*Phalaris arundincea* L.) foliage was considerably less than predicted from the azalea model (McCrady and Maggard 1993). The lower-than-predicted concentrations on grass were attributed to two factors: the difference in species characteristics, and provision for exposure to sunlight, which demonstrated that photodegradation was a significant dissipation mechanism. Photodegradation produced an approximate threefold reduction in the predicted equilibrium concentration compared with the predicted concentration in dark controls.

Similar studies to characterize other important agronomic species used for food and feed are required for quantitative assessment of transport of chemicals through food chains.

The adsorption of vapors to plants is not only important in evaluating aerial transport of nonpoint source contamination, but it also is an important factor in the transport of contaminants to plants in localized situations in which soils serve as a reservoir of residues. The volatilization–redeposition pathway has not been properly evaluated in many plant uptake studies, but several adequately designed studies demonstrated that this is the sole pathway for foliar contamination of plants grown in media containing lipophilic compounds (Bacci et al. 1992; Beall and Nash 1971; Fries and Marrow 1981; McCrady et al. 1990). Modeling this pathway in situations of localized contamination will be more complex than in the air–plant pathway of widespread contamination because the dynamics of air movement will make predictions of volatilization rates and air concentrations difficult.

The findings of several field studies involving localized contamination are qualitatively consistent with laboratory findings and modeling. Residues of 1,1-dichloro-2,2-*bis*-(*p*-chlorophenyl)-ethylene (DDE) in milk of cows grazing in pastures with contaminated soil were greatest when soil moisture and temperature conditions would tend to favor DDE volatilization (Willett et al. 1993). Concentrations of 1,1,1-trichloro-2,2-*bis*(4-chlorophenyl)-ethane (DDT) in soil were greater than DDE, but DDT residues were not detected in grass or milk. This finding can be attributed to the eightfold higher vapor pressure of DDE (Suntio et al. 1988). Residues were not detected in forage harvested from fields contaminated with PBBs (Fries and Jacobs 1986), and concentrations of PCDDs and PCDFs (mainly octachloro), in hay were not related to soil concentrations (Hulster and Marschner 1993). These observations are consistent with the relatively low vapor pressures of the PBBs (Fries 1985b) and octachlorodioxin (Shiu et al. 1988).

C. Particle Deposition and Retention on Plants

The interception and retention of dry deposition on plants has been studied and modeled extensively with respect to radioactive fallout (Baes et al. 1984). The interception fraction, which is the fraction of particles initially deposited and retained on the plants, is related to leaf area and roughness, plant biomass, plant density, and particle size (Martin 1964; Witherspoon and Taylor 1970). Increasing particle size generally decreases the likelihood of initial retention. Interception fractions as functions of dry matter yield have been derived for the major classes of food and feed crops (Baes et al. 1984). Removal or reduction in concentration of intercepted particles on plants is accomplished by weathering (wind action, precipitation washoff), growth dilution, and grazing by animals. Weathering losses tend to be rapid in the first few days following deposition, followed by a slower exponential (Martin 1964; Witherspoon and Taylor 1970). Half-lives of particle fallout on plants measured in field studies are in the range of 2 to 34 d, with a median value of 10 d, but a conservative value of 14 d is commonly assumed in risk assessments (Baes et al. 1984). In contrast, leaf washoff in some grasses involves movement of particles to the plant base, providing continued accessibility of contaminants to grazing animals (Russell 1963).

The 14-d value for the weathering half-life often is assumed to be short compared with the lengths of the growing period for many plants (Moghissi et al. 1980). Thus, the concentration of contaminants on plant foliage would exceed 90% of the steady-state concentrations in 8 wk if the deposition rate has been uniform. Concentrations in plants can be described by the equation

$$C = I F(1 - e^{-kt})/k Y, \qquad (2)$$

where C (pg/g) is concentration at time t (days), I is the interception fraction, F (pg/m^2/d) is the deposition flux, k (day^{-1}) is the weathering constant (0.0495 with a 14-d half-life), and Y (g/m^2) is the dry matter yield. As time increases, the exponential term approaches zero and the steady-state concentration (C_s) is

$$C_s = 20.2 \ I F/Y. \qquad (3)$$

Concentrations on plants predicted by Eq. (3) are expected to be overestimates because of the conservative 14-d half-life for weathering loss and because the constant k accounts for weathering loss but not for dilution during growth. This model was developed for inorganic chemicals that are not expected to partition from the particle to other phases or be subjected to dissipative processes. The organic chemicals that are of greatest importance as environmental pollutants interchange between the particle to vapor phases with changes in ambient temperature so that loss by volatilization or partitioning to the lipophilic cuticle waxes might be expected. The role of photodegradation is also uncertain in these circumstances, in which compounds are adsorbed to particles and may be protected from light.

D. Wet Deposition on Plants

Wet deposition involves the washout of particles and entrained vapors by precipitation and may account for more than one-half of the total deposition of airborne contaminants (Koester and Hites 1992). Precipitation reaching the top of a crop canopy may be partitioned into throughfall, stemflow, and retained on the plant (Paltineanu and Apostol 1974; Steiner et al. 1983). Throughfall is that portion of the precipitation that is not intercepted by the plant leaf or, if intercepted, is immediately lost to the soil by splashing off the leaves. Stemflow is that portion of the precipitation that is intercepted by the leaves and flows to and down the stem to the soil. The amount of precipitation retained on the plant is the difference between that measured at the top of the plant canopy and that measured as throughfall and stemflow. The amount retained on the plant usually is small compared with the total, and because it is measured by difference, the measurements have a high relative variation. Throughfall in closed corn (*Zea mays*) canopies is in the range of 40–50% of the rainfall, whereas throughfall in soybean (*Glycine max*) was more than 75% (Dowdy et al. 1993; Steiner et al. 1983). Throughfall may be assumed to make no direct contribution to plant contamination. The remaining precipitation would be intercepted by the plant and lost as stemflow, or retained on the plant, where it would be subject to evaporation after precipitation. The water-holding capacity of a given plant may be assumed to be finite and the maximum holding capacity reached after a given amount of precipitation. In corn with a closed canopy, 2 to 3 mm were retained when precipitation per event was 25 to 30 mm (Steiner et al. 1983).

Comparable amounts were found in another study when precipitation was 7.5 mm or more, but retention was less than this when precipitation was less than 5 mm (Parkin and Codling 1990). It is reasonable to assume that entrained vapors in the moisture remaining on the plant would be partitioned to cuticle waxes or lost by volatilization, whereas particles would remain on the plant and have the same fate as particles from dry deposition. The fate of contaminants in stemflow is more difficult to predict and depends on whether contact time is sufficient for compounds to partition to cuticle waxes as water flows over the plant surfaces. To some degree, particulates in stemflow would be expected to accumulate in the leaf-stem nodes and the base parts of the plants, as has been noted for washoff of dry deposition. Even if compounds from wet deposition are partitioned to cuticle waxes, desorption may also be expected so that the ultimate plant concentrations would be concentrations that are in equilibrium with ambient air. It is concluded that wet deposition will not be a significant factor in plant contamination by volatile compounds and is only of small importance in the case of compounds adsorbed to particles.

E. Other Deposition Processes

Localized contamination situations often involve deposition of chemicals in matrices that may significantly alter the retention and fate of the chemicals on plants. Many of the halogenated aromatic contaminants are photodegraded rapidly by the mechanism of reductive dehalogenation (Crosby and Wong 1977; Hutzinger et al. 1972; Ruzo et al. 1976). Thus, if the matrix includes materials that serve as hydrogen donors, enhanced degradation of a contaminant might be expected when exposed to sunlight. For example, pure TCDD is quite stable in sunlight, but loss from glass and leaf surfaces was rapid when TCDD was applied with herbicide formulating agents (Crosby and Wong 1977). Formulating agents also affect the volatilization of compounds from plant and soil surfaces. Application of TCDD with an emulsifying agent resulted in greater initial retention on plants and loss by photodegradation and volatilization than application in a granular formulation (Nash and Beall 1980). A study of TCDD residues in grass treated with contaminated phenoxy herbicides confirmed the rapid loss of TCDD initially, but a small fraction of the initial concentration was quite persistent (Jensen et al. 1983). Although the presence of TCDD in phenoxy herbicides is no longer an important concern, it can be concluded that the matrix material present when a volatile or photolabile compound is released to the environment may alter its fate significantly.

Application of sewage sludge to established crops may present a situation in which retention of semivolatile and photolabile compounds is enhanced. Liquid sludge applied to tall fescue (*Festuca arundinacea*) constituted approximately 30% of the forage dry matter immediately after application (Chaney and Lloyd 1979). Sludge allowed to dry for 24 hr was not removed from plants by rainfall or other physical processes, and the heavy metals in the sludge became incorporated into the plant (Chaney and Lloyd 1979; Jones et al. 1979). Data on volatilization losses of sludge-borne organics on plant leaves are not available, but adsorption of PCBs to sludge and other adsorbents has reduced volatility losses from soil, and this mechanism also might occur with sludge-borne contaminants on leaves (Fairbanks et al. 1987; Strek et al. 1981). A reduction in the rate of photolysis also might be expected because light would not penetrate the sludge.

F. Accumulation and Fate in Soil

The fraction of both dry and wet deposition from a continuing source that is initially intercepted by plants is appreciable, but most contaminants eventually reach soil by wind action or precipitation washoff from the plants. Because interception fractions never exceed 1.0, it follows from Eq. (3) that the maximum amount of a compound that can be removed from an area with a single crop is 20 times the daily flux, or approximately 5% of the yearly flux, which can be disregarded when predicting steady-state concentrations in soil. The concentration of a contaminant that would accumulate in soil as the result of a continuing input can be calculated by the equation

$$C = F(1 - e^{-kt})/kdD, \tag{4}$$

where C (pg/g) is concentration at time t (days), F (pg/m^2/day) is the flux, k (day^{-1}) is the dissipation rate constant, d (g/m^3) is the density of soil, and D (m) is depth of mixing. The commonly used value for the density of soil is 1500 kg/m^3, which is equivalent to 225 kg/m^2 in the tillage layer (top 15 cm) or 15 kg/m^2 for each cm of depth. At steady state, the exponential term approaches 0 and concentration is

$$C = F/kdD. \tag{5}$$

Essentially, the predicted concentration is directly proportional to the flux and inversely proportional to the dissipation rate and depth of mixing.

This description is an oversimplification because there are a number of dissipation processes that can be modified by environmental factors, with the result that dissipation in the environment rarely follows first-order kinetics.

The major dissipation processes are leaching, photodegradation, volatilization, and biodegradation. The nonpolar contaminants are strongly adsorbed to organic matter, and leaching generally is not an important factor affecting movement and dissipation of these compounds in soil (Filonow et al. 1976; Jackson et al. 1985; Paustenbach et al. 1992; Wild et al. 1991). The presence of such cocontaminants as chlorophenols that enhance the water solubility, or oils that serve as carriers, have in some instances led to leaching of nonpolar compounds (Freeman and Schroy 1989; Kapila et al. 1989). Photodegradation may occur when compounds are applied to the soil surface, and this process, in conjunction with volatilization, may lead to the rapid initial dissipation of compounds such as TCDD that are released into the environment at the soil surface (Crosby and Wong 1977; Fanelli et al. 1982; Kearney et al. 1972). Photodegradation is limited to a very shallow depth, which was between 0.06 and 0.13 mm in studies with octachlorodibenzo-p-dioxin (Miller et al. 1989).

Volatilization is the most significant dissipation process for compounds that are resistant to microbial degradation. Because it may be assumed that concentrations in surface soil will be in equilibrium with concentrations in ambient air under conditions of continuing nonpoint-source contamination, volatilization is mainly important in situations that involve single or episodic introductions of a contaminant in localized environments. Volatilization occurs at the soil–air interface, and it will impact soil concentrations to the depths to which compounds may be transported to or from the soil surface in the vapor phase through the unsaturated zone (Freeman and Schroy 1989; Jury et al. 1983). Thus, a typical scenario is illustrated by TCDD in the Seveso event in which the TCDD deposited at the soil surface was lost at a rapid rate initially and, after a year, greater concentrations were found at the 0.5–1.5 cm depth than at the 0–0.5 cm depth (Fanelli et al. 1982). The contaminant will be distributed uniformly in the tillage layer in cultivated soils, but if tillage ceases, loss of compound would be greater near the soil surface. Thus, residues of organochlorine insecticides in soils that were not cultivated for 13 yr were greater at depths of 15–20 cm than at depths near the soil surface (Nash and Woolson 1968). The rate of volatilization is directly related to vapor pressure, which within a class of compounds such as the PCDDs and PCBs is inversely related to molecular weight or degree of chlorination (Shui et al. 1988; Shui and Mackay 1986). Because vapor pressure is related to temperature, large diurnal and seasonal fluctuations in volatilization losses are expected (Freeman and Schroy 1989). The organic matter content is an important factor in determining volatilization losses of nonpolar compounds because these compounds are adsorbed to organic matter of soils (Haque et al. 1974). Thus, amendment of soil with organic matter in the form of sewage sludge and activated carbon, a strong adsorbent, significantly reduced the volatilization losses of PCBs (Fairbanks et al. 1987; Strek et al. 1981).

Microbial degradation is the only significant means of dissipation of nonpolar compounds from soil below the shallow surface areas where volatilization and photodegradation processes are active. The susceptibility of chemicals to microbial degradation is directly related to structure. Typically, the metabolism of aromatic hydrocarbons initially involves oxidation to a diol and subsequent ring cleavage (Fishbein 1984). Because this pathway requires two adjacent nonsubstituted carbons, the degree and position of halogen substitution on the ring is the important determinant of the persistence of a halogenated aromatic. Most PCBs with four or fewer chlorines are readily metabolized by microorganisms, with the result that most congeners found in environmental samples contain six or more chlorines (Furukawa and Matsumura 1979, Furukawa et al. 1979). If the microbial degradation of the more chlorinated compounds occurs, the rate is slow and of little practical significance. An analogous situation exists with PCDDs and PCDFs, and no mechanisms for the biological degradation of these compounds chlorinated in the 2, 3, 7, and 8 positions have been identified (Paustenbach et al. 1992). The PAHs are susceptible to degradation or incorporation into humic material, as measured by disappearance of the parent compound in soil (Bossert and Bartha 1986; Wild and Jones 1993). The rate of degradation is related to the number of aromatic rings and the log K_{ow}. The range of half-lives was as short as <2 yr for naphthalene to as great as 16.5 yr for coronene (Wild et al. 1991). Nitrosamines also appear to be resistant to degradation by soil microorganisms because of the apparent resistance of the nitrogen–nitrogen bond to microbial attack (Tate and Alexander 1976). The phthalate ester di-2-(ethylhexyl) phthalate (DEHP) had a 50-d half-life in soil, which was attributed to microbial degradation because volatilization loss was not detected (Fairbanks et al. 1985).

It follows from the preceding discussion that the rates of dissipation of recalcitrant compounds are dependent on the degree to which the compounds at the soil surface are diluted by tillage or other mechanical actions. However, the negative impact on dissipation is mitigated by reduction in the amounts of compound that can be transported into other segments of the food chain by processes that are dependent on concentrations at the soil surface.

G. Transport from Soil to Plants

A comprehensive review on the uptake of environmental contaminants by plants is available (O'Connor et al. 1991). Residues are transported from soil to aerial parts of plants primarily by two mechanisms: (1) root uptake and translocation within the plants and (2) volatilization from the soil surface and redeposition on plants. The relative significance of the two processes depends upon the water solubility of the compound (Ryan et al. 1988). Compounds with low log K_{ow}s of approximately 10^{-2} are the most likely to be taken up by roots and translocated to aerial plant parts if the compound is not metabolized by the plant. When log K_{ow} is >5.0, the major transport mechanism would be volatilization and redeposition, and the quantity transported would depend on the vapor pressure and the Henry's law constant of the compound. An inverse relationship between log K_{ow} and log BCF (bioconcentration factor, C_{plant}/C_{soil}) was demonstrated in a compilation of literature values (Travis and Arms 1988). The existence of this relationship was probably fortuitous because log K_{ow} and vapor pressure often are inversely related.

The potential transfer of residues of persistent chlorinated compounds that are semivolatile, such as the PCDDs, PCDFs, and PCBs, is expected to be achieved by volatilization and redeposition because the log K_{ow}s of these compounds exceed 5.0. This mechanism has been confirmed in several studies in which no plant residues were detected when suitable vapor barriers were provided (Beall and Nash 1971; Fries and Marrow 1981; McCrady et al. 1990). The two steps in this mechanism of plant contamination, volatilization from soil and the air–plant exchange, have been discussed in previous sections. The volatilization-redeposition mechanism has important implications for evaluating the transport of contaminants into animal food chains. Only the concentrations of contaminants near the soil surface are important in determining plant contamination, and seeds used for feed generally will be contaminant-free because the compounds will not be translocated from the plant surface.

The relative vapor pressure of a homologous series of chlorinated compounds decreases with increasing chlorination, and little vapor transfer of the highly chlorinated compounds is expected. The only potential transport mechanism for the highly chlorinated congeners is by transfer of soil particles as dust or by splashing during precipitation. This mechanism does not appear to be particularly important for animal forage crops. Concentrations of PCDDs/PCDFs, mainly hepta and octa, in hay were unrelated to soil concentrations (Hulster and Marschner 1993), and no residues of PBB were detected in harvested forages grown in soil with concentrations as high as 300 ppb (Fries and Jacobs 1986).

The PAHs have log K_{ow} values that range from approximately 3.5 to 7.5 (Wild and Jones 1991). Thus, some of lower-molecular-weight PAHs may be taken up by the roots and translocated to other plant parts, as demonstrated in studies with ^{14}C-labeled compounds (Edwards 1983). There is some evidence of catabolism in the plant, but the pathways have not been defined. As molecular weight and log K_{ow} increase, the soil-to-plant pathway would tend to be volatilization and redeposition, and the transport would be subject to the same variables as described for halogenated compounds. Volatilization losses from soil are not significant for PAHs with four or more rings (Park et al. 1990). Thus, the concentrations of PAHs on most crops are independent of soil concentrations and are thought to arise from aerial deposition (Jones et al. 1989a; O'Connor et al. 1991).

Plant accumulation of the other contaminants does not appear to be an important transport mechanism. Phthalate esters were taken up by plants, as demonstrated in studies with ^{14}C-labeled DEHP, but the DEHP apparently was metabolized because the parent compound was not detected in the plant (Aranda et al. 1989). Nitrosamines and phenolic compounds also appear to be taken up and metabolized by plants (Kearney et al. 1980b; O'Connor et al. 1991).

IV. Animal Uptake and Disposition
A. Animal Management Systems and Exposure Pathways

The consumption of contaminated feed and soil are the only significant pathways of animal exposure to environmental contaminants. The relative importance of these two sources depends on the animal types and the farm management systems. The intake of fibrous feeds derived from plant stems and leaves is the most important factor determining animal exposure because feeds derived from seeds are not expected to contain persistent lipophilic compounds. Roughage can be the major or sole feed for ruminants, but concentrates derived from seeds are the major or sole feed for poultry and swine.

The use of pasture as a roughage source is of particular importance because it adds soil ingestion to direct plant contamination as a source of exposure. An additional significance of pasture is that some processing methods, such as drying hay, may reduce concentrations of contaminant by volatilization (Archer and Crosby 1969). Replacement dairy heifers, sheep, young beef cattle, and beef cows used for calf production often subsist on diets that are mainly forage, and these animals usually have access to pasture. However, lactating dairy cows, often receive <50% of their dry matter from concentrates, and beef cattle in the finishing stages before slaughter receive almost no forage (U.S. Department of Agriculture 1992). Most commercial poultry and swine production is conducted in confined operations, which eliminates contaminated soil as an exposure source.

Unrestricted feed intake of farm animals is closely related to body weight as well as the physiological processes of growth, lactation, and reproduction. Daily consumption of dry feed by adult ruminants is approximately 2% of body weight for nonlactating animals and 3.3% for lactating animals (Subcommittee on Feed Intake 1987). For example, a typical 500 kg lactating cow consumes approximately 16 kg/d of dry matter, but a nonlactating animal of the same weight will only consume 10 kg. Individual animals and circumstances may cause deviations from these averages, but the deviations usually are not great enough to impact exposure estimates significantly. However, forage intake of grazing animals is reduced greatly when the standing crop is < 1500 kg/ha (Subcommittee on Feed Intake 1987). This observation has important implications for estimating quantities of soil ingested by grazing animals.

B. Soil Ingestion

Soil ingestion by cattle, sheep, and swine has been measured in a number of geographical areas under a variety of conditions (Fries et al. 1982a,b; Healy 1968; Healy et al. 1967, 1970; Healy and Drew 1970; Healy and Ludwig 1965; Mayland et al. 1975; McGrath et al. 1982; Thornton and Abrahams 1981). All studies employed methodology in which a nonabsorbed indicator, either acid-insoluble ash or titanium oxide, was measured in feces. Because these substances are not contained in plants, it was assumed that the concentrations of indicators in feces were a direct measure of the concentrations of soil in the feces. A typical algorithm for estimating the amount of soil consumed is

$$I_{soil} = I_{DM}(1 - D)F_{soil}/(1 - DF_{soil}), \tag{6}$$

where I_{soil} is the amount of soil ingested (kg/d), I_{DM} is the dry matter consumed (kg/d), D is the assumed fraction of dry matter digested coefficient, and F_{soil} is the fraction of soil in feces (Fries et al. 1982a). It was necessary to assume values for digestibility and feed intake because these parameters are difficult to measure in grazing animals.

Generally, soil ingestion is related inversely to availability of forage when pasture is the sole source feed. Soil intakes as low as 1% or 2% of dry matter intake for cattle and sheep occurred in spring when grass was abundant. When forage was sparse during fall or winter, intake of soil was estimated to be as great as 18% of the diet. The average soil ingestion was approximately 6% of dry matter intake for dairy cattle and 4.5% of dry matter intake for sheep when pasture was the only feed source of animals grazing 365 d/yr in New Zealand (Healy 1968; Healy and Ludwig 1965). Thus, the quantity of soil ingested was approximately 900 g/d for a 500 kg dairy cow and 45 g/d for a 50 kg sheep.

The tendency to use conservative parameters in exposure assessments has led to the frequent use of the New Zealand findings (U.S. EPA 1989). In many circumstances, these values are unrealistically high. Climatic conditions that prevent year-round grazing would reduce the total amount of soil ingested, even though the extreme values at the beginning and end of the grazing season were similar to the New Zealand findings (Mayland et al. 1975; McGrath et al. 1982; Thornton and Abrahams 1981). Supplemental feeding of other roughages or concentrate feeds reduces average yearly soil ingestion values significantly (Healy et al. 1967; Healy and Drew 1970). This has important implications for lactating dairy cows that usually are fed a supplemental concentrate, and it is unlikely that soil ingestion would ever exceed 1% or 2% of dry matter intake.

Two factors that are commonly overlooked may have led to overly high estimates of soil ingestion during periods of sparse grazing. Single values for digestibility and dry matter intake have been used for the complete grazing cycle in studies of soil ingestion with only rare exceptions (Mayland et al. 1975). Digestibility decreases as plants mature, and use of a single digestibility value in Eq. (6) generally will overestimate digestibility and soil ingestion late in the season. The use of a single value for dry matter intake over the complete grazing season also may lead to overestimation of soil ingestion during periods of low forage availability. Maximum forage intake by cattle and sheep is attained when the standing forage crop exceeds 2250 kg/ha or the available organic matter exceeds 40 g/kg of animal live weight (Subcommittee on Feed Intake 1987). Dry matter intake progressively declines with reduced amounts of available forage. The only study of soil ingestion that included measurements of the amount of standing forage demonstrated marked increases in concentrations of fecal acid-insoluble ash when amounts of forage organic matter fell below 50 g/kg (McGrath et al. 1982). Estimates of the amount of soil ingested were reduced by as much as 50% when forage intake values were adjusted for the available forage (Fries 1994). Circumstances of low forage availability existed in most studies of soil ingestion, and it is probable that many of the higher values for soil ingestion in the literature are erroneously high because of failure to consider lowered forage intake when grazing was sparse.

There are several situations other than grazing in which soil ingestion can be the source of animal exposure and residues in products. Cattle confined in unpaved lots consume a small amount of soil, and it was demonstrated on farms contaminated with PBB that this soil could be a source of tissue and milk residues (Fries and Jacobs 1986; Fries et al. 1982a). Although most commercial swine and poultry production is conducted in confined operations, the exposure of these species can be significant when they have access to contaminated soil. Pigs may ingest as much as 8% of their dry matter intake as soil because of their rooting habits (Fries et al. 1982b). The greater relative soil consumption by pigs than cattle was reflected in higher tissue-to-soil concentration ratios in pigs than in cattle on PBB-contaminated farms (Fries and Jacobs 1986). Free-range chickens also appear to have relatively high soil ingestion rates (Petreas 1991).

C. Effects of Matrices on Bioavailability

Environmental contaminants are associated with a variety of matrices, such as plant material, soil, fly ash, and sewage sludge, that may affect bioavailability, which is defined here as the fraction of the compound that is absorbed and stored, metabolized, or excreted by the animal. True bioavailability can be evaluated only in mass balance experiments, which rarely have been conducted with environmental contaminants. Because recirculation to the gastrointestinal tract is a route of excretion for many lipophilic chemicals, it generally is necessary to treat the amount of compound recirculated and excreted via feces as unavailable. The degree of chlorination and other molecular characteristics also affect absorption and bioavailability as defined, but these effects will be discussed in a subsequent section on bioaccumulation.

Typically, laboratory studies of animal uptake have been conducted with compounds incorporated in normal diets or in a carrier, such as corn oil. Lipophilic compounds in oil solutions have a high bioavailability, and uptake may be as great as 90% of the ingested compound, but uptake is less when compounds are present in normal diets. For example, net absorption of TCDD by rats was 75–80% when administered in corn oil (Rose et al. 1976), but was only 50–55% when TCDD was contained in normal rat and cow diets (Fries and Marrow 1975a; Jones et al. 1989). Reviews of the available literature indicated uptake of TCDD from normal soils is approximately 40%, as low as 30% from fly ash, and <20% in soils from some industrial sites (Fries 1991; Fries and Paustenbach 1990). When TCDD was administered with activated carbon, a strong adsorbent, uptake by rats was negligible (Poiger and Schlatter 1980). Data are not available for such comprehensive comparisons for other lipophilic compounds, but comparable results can be anticipated.

Lipophilic compounds in soil tend to be adsorbed to the organic matter fraction, which varies greatly among soils and can be altered by sludge and other amendments (Burchill et al. 1981). Reduced mineralization rates of PCBs in sludge-amended soils were related to the increased adsorption of PCBs in soil when the organic matter was increased (Fairbanks and O'Connor 1984; Fairbanks et al. 1987). Some evidence suggests that uptake of halogenated hydrocarbons by earthworms is inversely related to the organic matter content of soil (Davis 1971), but the uptake of PCBs by rats was unaffected by the organic matter content (Fries and Marrow 1992). Amendment of soils with activated carbon, a strong adsorbent, has yielded comparable results. Plant uptake of PCBs from soil was reduced (Strek et al. 1981), but animal uptake of PBBs from soil was not affected significantly (Fries 1985b). The failure of differences in soil adsorption caused by organic matter content to be translated into differences in animal uptake may be related to the nature of the adsorption mechanism and physical chemical conditions of the gastrointestinal tract. If the adsorption of lipophilic compounds in soil is viewed as partitioning from the aqueous phase (Burchill et al. 1981), it follows that the reverse would occur in the presence of dietary lipids and bile salts in the gastrointestinal tract.

The adsorption of polar compounds to soil involves an electrostatic attraction to negatively charged clay and mineral particles and may differ by orders of magnitude depending upon the properties of the soil and/or the compounds (Weber 1972). These differences in adsorption may be related to biological effects, such as phytotoxicity (Harrison et al. 1976), but no information was found relating adsorption to bioavailability of the compounds in animals. The acidic conditions of parts of the gastrointestinal tract may lead to the dissociation of chemicals that are adsorbed at the pH range of normal soils and thereby make the compounds available for absorption. If desorption occurred in the gastrointestinal tract, it would be of little significance in the movement of compounds through the food chain because polar compounds are readily metabolized and excreted in the urine by animals.

D. Bioaccumulation of Halogenated Aromatics

The traditional approach for measuring and comparing accumulation of persistent compounds in animals and animal products is the bioconcentration factor (BCF), which is the ratio of the concentration in the animal tissue or product to the concentration in the diet (Kenaga 1980; Fries 1991; Fries and Paustenbach 1990). Another measure, used less frequently, is the biotransfer factor (BTF), which is the ratio of the concentration in the product to the amount of compound ingested (Travis and Arms 1988). The conclusions drawn from both approaches are interchangeable because both are derived from the same databases and incorporate similar assumptions concerning feed intake or animal productivity.

Ideally, BCFs and BTFs should be determined at steady state when the rate of ingestion is equal to the rate of metabolism and excretion. Concentrations in milk fat usually attain an approximate steady state within 40–60 d of continuous feeding (Fries 1977), but concentrations in body fat require 6 mon or longer to reach equilibrium (Fries 1982, 1991). This difference in time to reach equilibrium probably suggests a greater ease of transport of lipophilic chemicals across membranes in the mammary gland than in adipose tissues. As a result, concentrations in milk fat are closely related to concentrations in blood, which often reflect short-term changes in concentration in the gut pool. Many published studies are too short to reach steady state, particularly in body fat. Additionally, the concept of steady state requires that contaminant concentrations in diets, body fat, pool sizes, and elimination rates remain constant, a condition that rarely occurs in practice. Application of pharmacokinetic models would be an approach for estimating residue concentrations at steady state, but models incorporating all of the enumerated variables have not been developed. A final caution in the use of BCF data is that analytical and quantification methods have changed significantly over the years since some of the data were collected.

Despite the deficiencies of individual experiments, data on BCFs of halogenated hydrocarbons are consistent when examined as a whole. Several compilations of BCFs in milk fat are available (Fries 1991; Noble 1990; Travis and Arms 1988). Accumulation of persistent organics in meat animals has not been studied as extensively as in dairy animals, and many studies with meat animals are shorter than the time required to reach steady-state concentrations. Several longer-term studies that included re-peat tissue samples indicated that constant concentrations of several chlorinated hydrocarbons were reached within 100–200 d, and a compilation of results of studies of this length is available (Fries 1991).

The maximum BCFs for halogenated hydrocarbons were in the range of 5–6 for both milk fat and body fat, and the ranges were similar in lactating and nonlactating animals. Superficially, lower BCFs in lactating animals might be expected because of the importance of milk fat as a route of elimination of lipophilic chemicals. However, lactating animals consume more feed per unit body weight (Subcommittee an Feed Intake 1987), which may account for the similarity in BCFs.

The BCFs of many halogenated hydrocarbons are lower than the five to six maximum values and, in many instances, there is no evidence for bioaccumulation. This range in BCFs is illustrated with individual PCB congener measurements made in the milk of cows after 60 d of continuous dosing of Aroclor 1260 and several mono- and dichloro-PCB congeners (Table 1). The congeners may be separated into four groups: those with no residue in milk, and those with low, intermediate, and high BCFs. All congeners with high BCFs had six or more chlorines, and chlorine substitutions in the 4,4′ positions (Figure 1), an observation that was noted by McLachlan (1993). The congeners that were completely metabolized or had low BCFs lacked the 4,4′ substitution and had two adjacent nonsubstituted positions on at least one ring. No consistent substitution pattern can be discerned in the three congeners with intermediate BCFs. A comparable situation exists with the PCDDs and PCDFs in which chlorination in the 2,3,7,8 positions and 1,2,3,7,8 positions (Figure 1), respectively, inhibit metabolism (McLachlan et al. 1989; Olling et al. 1991).

The BCFs of nonmetabolized PCB, PBB, PCDD, and PCDF congeners decrease as the degree of halogenation increases (Fries and Marrow 1975; McLachlan 1993; Tuinstra et al. 1992). The only consistent interpretation would be that increased halogenation decreases rates of absorption from the gastrointestinal tract and across other biological membranes. This interpretation is supported by limited mass balance data in cows in which the fecal excretion of the compounds with the low BCFs was enhanced relative to compounds with higher BCFs (Willett and Durst 1978; McLachlan 1993; McLachlan et al. 1989).

The use of BCFs, or similar factors, to express transfer of compounds from diet to products implicitly assumes that these factors are independent of the dietary concentration or dose rate. An analysis of all available data for the PCB Aroclor 1254 suggests that the BCFs were not constant, and the concentration of PCB in milk fat was best described as a function of daily dose (mg/kg body weight) raised to the 0.81 power (Willett et al. 1990). At any given level of feed intake, application of this function would predict lowered BCFs with increased dose rates or concentrations in the diet.

Table 1. Polychlorinated biphenyl congeners falling within specified ranges of biological concentration factors (BCFs) for milk fat in dairy cows.

Number	Chlorines			Substitution pattern								log K_{ow}	BCF
		2	3	4	5	6	2'	3'	4'	5'	6'		
No Accumulation													
3	1			4								4.5	ND
4	2	2					2'					4.9	ND
87	5	2	3	4			2'			5'		6.5	ND
97	5	2		4	5		2'	3'				6.6	ND
136	6	2	3			6	2'	3'			6'	6.7	ND
185	7	2	3	4	5	6	2'			5'		7.0	ND
Low BCF													
7	2	2		4								5.0	0.1
15	2			4					4'			5.3	0.1
1	1	2										4.3	0.4
101	5	2		4	5		2'			5'		6.4	0.4
149	6	2	3			6	2'		4'	5'		6.8	0.4
151	6	2	3		5	6	2'			5'		−	0.4
95	5	2	3			6	2'			5'		6.4	0.5
141	6	2	3	4	5		2'			5'		−	0.5
Intermediate BCF													
187	7	2	3		5	6	2'		4'	5'		−	1.0
198	8	2	3	4	5		2'	3'		5'	6'	−	1.4
52	4	2			5		2'			5'		6.1	1.8
High BCF													
138	6	2	3	4			2'		4'	5'		7.0	3.5
170	7	2	3	4	5		2'	3'	4'			−	3.6
194	8	2	3	4	5		2'	3'	4'	5'		7.1	3.7
180	7	2	3	4	5		2'		4'	5'		−	4.0
206	9	2	3	4	5	6	2'	3'	4'	5'		7.2	4.0
128	6	2	3	4			2'	3'	4'			7.0	4.1
153	6	2		4	5		2'		4'	5'		6.9	4.5

BCFs were calculated from data in Tuinstra et al. (1981), log K_{ow}s from Shiu and MacKay (1986), and congener numbers from Ballschmiter and Zell (1980).

E. Fate of Other Contaminants

Little pharmacokinetic information is available on the PAHs, phthalates, nitrosamines, phenolics, and other nonhalogenated environmental contaminants. These compounds rarely have been identified in animal products and, thus, the need to obtain pharmacokinetic information has not received high priority. The available information, including data on the physical-chemical properties of the compounds, suggests that the likelihood of transmission to food products is low.

The PAHs are metabolized by epoxidation, hydroxylation, and conjugation to water-soluble compounds that are excreted by the animal (Zedeck 1980). Studies in rats using single oral doses of several PAHs indicate that metabolic degradation of PAHs was described by a two-compartment first-order model (Withey et al. 1991; Modica et al. 1983). The rates of metabolism were rapid, and over 80% of the radioactivity from a single dose of pyrene was excreted within 2 d (Withey et al. 1991). Small amounts were stored in tissues with the highest concentrations in fat, but the concentrations are such that significant bioaccumulation would not be expected. Radiolabeled naphthalene has been administered to several farm animal species in single and 31-d continuous dose studies (Eisele 1985). Concentrations of radioactivity in the fat and the aqueous portions of milk were approximately equal to steady-state concentrations in each fraction approximating $20\text{--}25 \times 10^{-6}\%$ dose/g, which is equivalent to a BCF of 0.32×10^{-6}. The concentration in body fat at 3 d after dosing was only 0.001% dose/g, which also indicates little bioaccumulation.

General surveys also support the conclusion that the likelihood of transmission of PAHs to animal food products is low. Cattle and pig tissues from areas of high and low environmental levels of PAHs in feeds were compared and benzo(a)pyrene was not detected at levels above 0.05 ppb in either case (Lusky et al. 1992). Concentrations of PAHs in dairy products were lower than the concentrations of PAHs in vegetable products with similar fat contents in surveys of retail food (Dennis et al. 1991), which indicates that animal metabolism reduces the transmission of PAHs to humans.

The transport of nitrosamines from the diet to milk of cows has been of interest because several of these compounds can be formed during fermentation of silage. No residues were found in milk when silages containing as much as 100 ppb dimethyl-, diethyl-, dipropyl-, or dibutylnitrosamines were fed to cows (Mohler and El-Refai 1981; Terplan et al. 1978, 1980). These results and the physical–chemical properties of nitrosamines make it unlikely that significant quantities of these compounds would be transmitted into human food products through animal food chains.

Phthalate esters are lipophilic and are readily absorbed and metabolized by animals (Kluwe 1982; Thomas and Northrup 1982). For example, more than 90% of single oral doses of di-n-butyl phthalate and di(2-ethylhexyl) phthalate (DEHP) were excreted in the urine of rats within 48 hr (Kluwe 1982). The esters were metabolized by hydrolysis of the side chains (Figure 3), with the formation of water-soluble products. Steady-state concentrations in fat of rats fed diets containing DEHP were approximately 8 ppm in animals fed 1000 ppm and 70 ppm in animals fed 5000 ppm (Daniel and Bratt 1974). Transfer of DEHP to milk has been demonstrated in lactating rats that received oral doses of 2000 mg/kg daily (Parmar et al. 1985). No reports were found that measured accumulation and fate in food animal species, but the findings in laboratory species suggest that there would be little transfer of phthalate esters to animal food products. Market milk surveys revealed rare occurrences of phthalate esters, and these rare occurrences were attributed to exposure of the product to phthalate-containing plastics used in processing and storage (Peterson 1991).

The chlorophenols have been studied extensively in livestock because of the use of treated wood and shavings for livestock shelter and bedding. This exposure has led to low concentrations of chlorophenols in animal tissues and products (Butler and Frank 1991; Frank et al. 1982; Osweiler et al. 1983; Ryan et al. 1985). As is typical of other polar compounds, tissue residue concentrations are highest in liver and kidney (Hughs et al. 1985, Kinzell et al. 1985). Radiolabeled pentachlorophenol is metabolized rapidly in the cow, with an elimination half-life of approximately 43 hr. Approximately 5% of the radiolabeled material was transferred to milk in both free and conjugated forms. Much of the pentachlorophenol is excreted in the conjugated form in the urine. The data suggest that there is little chance of accumulation of pentachlorophenol or other phenolic compounds.

This brief overview of the fate of several nonhalogenated lipophilic and water-soluble compounds is probably representative of the fates of the large number of similar compounds that might be introduced into the environment. It is concluded that the likelihood of transmission of compounds with these characteristics to foods is small because animal metabolism degrades most compounds before storage in tissues or excretion in products. Therefore, passage of these compounds through animal production systems would tend to reduce human exposures.

F. Prediction of Bioaccumulation

It is not practical to obtain BCF data for all compounds and animal species that may be involved in a specific situation of environmental contamination and it would be useful to be able to predict BCFs from the physical-chemical properties of organic compounds. The relationship between bioaccumulation (log BCF) and the octanol–water partition coefficient (log K_{ow}) has been the property used most frequently (Kenaga 1980; Travis and Arms 1988). Typically, the R^2s for the relationships in milk and tissue fats were in the range of 0.50–0.65. The variation not accounted for by K_{ow} appears to be related to the failure to distinguish between metabolically labile and nonlabile compounds, and a reduced rate of transfer of higher-molecular-weight compounds across biological membranes.

The deficiencies in the use of K_{ow} to predict BCFs is illustrated by the individual PCB congener measurements presented previously (Table 1). Congeners with the same number of chlorines have comparable K_{ow}s (Shiu and Mackay 1986), but the BCFs of congeners with the same degree of chlorination differed markedly with changes in substitution pattern, which is a factor in determining susceptibility to degradation. Another example is provided by the PAHs, which have high K_{ow}s (Wild and Jones 1991), but exhibit no bioaccumulation in animals. The second area in which the K_{ow}s do not provide adequate prediction is with the reduction in BCF values with increased chlorination of compounds like the PCDDs and PCDFs. McLachlan (1993) documented this phenomenon with PCBs and chlorinated hydrocarbons in dairy cattle and suggested that the decrease in BCFs began at log K_{ow}s of approximately 6.5. Because log K_{ow}s of PCBs are related closely to molecular weight (Shiu and Mackay 1986), the decline in absorption is not necessarily related to lipophilicity, but it could be related to some other parameter that is related to molecular weight. It is concluded that K_{ow} may be a helpful guide if no other data are available, but caution should be used in applying the predictions.

Summary

A large number of chemical contaminants potentially may be present in agricultural environments, leading to exposure of animals and potential residues in animal products. The contamination may be either widespread, as a result of aerial transport of industrial emissions, or localized, as a result of accidental emissions and spills, improper waste disposal, contaminants in useful products, and areas of past use of products now banned.

The halogenated hydrocarbons, including the polychlorinated dibenzo-p-dioxins (PCDDs), polychlorinated dibenzofurans (PCDFs), polychlorinated biphenyls (PCBs), and persistent organochlorine insecticides remaining from past use, are the contaminants of most concern. Depending on the degree and pattern of chlorine substitution, these compounds are resistant to degradation and tend to accumulate in the fat of animals and their products. Other classes of environmental contaminants as exemplified by the PAHs, phthalate esters, acid phenolics, and nitrosamines also may occur widely in the environment. These compounds are unlikely to be transported to animal products because the compounds are water-soluble or can be metabolized to water-soluble products, which are excreted in the urine and thus do not bioaccumulate in products such as milk and meat.

The points of entry of environmental contaminants into agricultural environments usually are plants and soils. Lipophilic compounds such as the halogenated hydrocarbons are not taken up and translocated by plants. Contamination of plants is mainly a surface phenomenon resulting from aerial deposition of emissions or deposition of compounds volatilized from the surface of contaminated soil. Thus, fibrous roughages used primarily in feeding cattle and other ruminants will be the most important pathway of animal exposure and transport to human foods. The second pathway of animal exposure is by ingestion of contaminated soil while grazing or when confined to unpaved facilities. As in the case of feed sources, cattle is the species most vulnerable to exposure by the soil ingestion pathway under most commercial management systems, but poultry and swine are more vulnerable in those infrequent situations in which these species have access to contaminated soil.

References

Aranada J, O'Connor GA, Eiceman GA (1989) Effects of sewage sludge on DEHP uptake by plants. J Environ Qual 18:45–50.

Archer TE, Crosby DG (1969) Removal of DDT and related chlorinated hydrocarbon residues from alfalfa hay. J Agric Food Chem 16:623–626.

Bacci E, Calamari C, Gaggi C, Vighi M (1990a) Bioconcentration of organic chemical vapors in plant leaves: Experimental measurements and corrolation. Environ Sci Toxicol 24:885–889.

Bacci E, Cerejeira MJ, Gaggi C, Chemello G, Calamari D, Vighi M (1990b) Bioconcentration of organic chemical vapors in plant leaves: The azalea model. Chemosphere 21:525–535.

Bacci E, Cerejeira MJ, Gaggi C, Chemello G, Calamari D, Vighi M (1992) Chlorinated dioxins: Volatilization from soils and bioconcentration in plant leaves. Bull Environ Contam Toxicol 48:401–408.

Baes CF, Sharp RD, Sjoreen AL, Shor RW (1984) A Review and Analysis of Parameters for Assessing Transport of Environmentally Released Radionuclides through Agriculture. ORNL-5786, Oak Ridge National Laboratory, Oak Ridge, TN.

Ballschmiter K, Zell M (1980) Analysis of polychlorinated biphenyls (PCB) by glass capillary gas chromatography. Fresenius Z Anal Chem 302:20–31.

Beall ML, Nash RG (1971) Organochlorine insecticide residues in soybean plant tops: root vs. vapor sorption. Agron J 63:460–464.

Bidleman TE (1988) Atmospheric processes: Wet and dry deposition of organic compounds are controlled by their vapor-particle partitioning. Environ Sci Technol 22:368.

Bilawchuk MS, Kitts DD, Owen BD (1989) Chemical characterization and toxicological assessment of kraft pulp mill fiber waste as a feedstuff for beef cattle. Bull Environ Contam Toxicol 43:52–59.

Bossert ID, Bartha R (1986) Structure-biodegradability relationships of polycyclic aromatic hydrocarbons in soil. Bull Environ Contam Toxicol 37:490–495.

Buckley EH (1982) Accumulation of airborne polychlorinated biphenyls in foliage. Science 216:520–522.

Bumb RR, Crummett WB, Cutie SS, Gledhill JR, Hummel RH, Kagel RO, Lamparski LL, Louma EV, Miller DL, Nestrick TJ, Shadoff LA, Stehl RH, Woods JS (1980) Trace chemistries of fire: A source of chlorinated dioxins. Science 210:385–390.

Burchill S, Greenland DJ, Hayes MHB (1981) Absorption of organic molecules. In: Greenland DJ, Hayes MHB (eds) The Chemistry of Soil Processes. John Wiley & Sons, Chichester, pp 621–670.

Butler KM, Frank R (1991) Pentachlorophenol residues in porcine tissue following preslaughter exposure to treated wood shavings. J Food Prot 54:448–450.

Carter CD, Kimbrough RD, Liddle JA, Cline RF, Zack MM, Barthel WF, Koehler RE, Phillips PE (1975) Tetrachlorodibenzodioxin: An accidental poisoning episode in horse arenas. Science 188:738–740.

Chaney RL, Lloyd CA (1979) Adherence of spray-applied liquid digested sewage sludge to tall fescue. J Environ Qual 8:407–441.

Cirelli DP (1978) Patterns of pentachlorophenol usage in the United States of America – An overview. In: Rao R (ed) Pentachlorophenols. Plenum Press, New York, pp 13–18.

Colborn T, vom Saal FS, Soto AM (1993) Developmental effects of endocrine-disrupting chemicals in chemicals in wildlife and humans. Environ Hlth Perspect 101:378–384.

Connett P, Webster T (1987) An estimation of the relative human exposure to 2,3,7,8-TCDD emissions via inhalation and ingestion of cow's milk. Chemosphere 16:2079–2084.

Crosby DG, Wong AS (1977) Environmental degradation of 2,3,7,8-tetra-chlorodibenzo-p-dioxin (TCDD). Science 195:1337–1338.

Czuczwa JM, McVeety BD, Hites RA (1984) Polychlorinated dibenzo-p-dioxins and dibenzofurans in sediments form Siskiwit Lake, Isle Royale. Science 226:568–569.

Daniel JW, Bratt H (1974) The absorption, metabolism and tissue distribution of di(2-ethylhexyl) phthalate in rats. Toxicology 2:51–59.

Davis BNK (1971) Laboratory studies on the uptake of dieldrin and DDT by earthworms. Soil Biol Biochem 3:221–233.

Davis DL, Bradlow HL, Wolff M, Woodruff T, Hoel DG, Anton-Culver H (1993) Medical hypothesis: Xenoestrogens as preventable causes of breast cancer. Environ Hlth Perspect 101:372–377.

Dean RB, Suess MJ (1985) The risk to health of chemicals in sewage sludge applied to land. Waste Mgt Res 3:251–278.

Dennis MJ, Massey RC, Cripps G, Venn I, Howarth N, Lee G (1991) Factors affecting the polycyclic aromatic hydrocarbon content of cereals, fats and other food products. Food Addit Contam 8:517–530.

Donnelly JR, Grange AH, Nunn NJ, Sovocool GW, Breen JJ (1990) Bromo- and bromochloro-dibenzo-p-dioxins and dibenzofurans in the environment. Chemosphere 20:1423–1430.

Dowdy RH, Lamb JA, Anderson JL, Alessi RS, Reicosky DC (1993) Rainfall distribution under corn and soybean canopies at the Minnesota MSEA site. In: Proceedings of the Conference on Agricultural Research to Protect Water Quality. Soil and Water Conservation Society, Ankeny, IA, pp 382–385.

Edwards NT (1983) Polycyclic aromatic hydrocarbons (PAHs) in the terrestrial environment – A review. J Environ Qual 12:427–441.

Eisele GR (1985) Naphthalene distribution in tissues of laying pullets, swine, and dairy cattle. Bull Environ Contam Toxicol 34:549–556.

Eitzer BD, Hites RA (1989a) Polychlorinated dibenzo-p-dioxins and dibenzofurans in the ambient air of Bloomington, Indiana. Environ Sci Technol 23:1389–1395.

Eitzer BD, Hites RA (1989b) Atmospheric transport and deposition of polychlorinated dibenzo-p-dioxins and dibenzofurans in the ambient air of Bloomington, Indiana. Environ Sci Technol 23:1396–1401.

Fairbanks BC, O'Connor GA (1984) Effect of sewage sludge on the adsorption of polychlorinated biphenyls by three New Mexico soils. J Environ Qual 13:297–300.

Fairbanks BC, O'Connor GA, Smith SE (1985) Fate of di-2-(ethylhexyl) phthalate in three sludge-amended New Mexico soils. J Environ Qual 14:479–483.

Fairbanks BC, O'Connor GA, Smith SE (1987) Mineralization and volatilization of polychlorinated biphenyls in sludge-amended soils. J Environ Qual 16:18–25.

Fanelli R, Chiabrando C, Bonaccorsi A (1982) TCDD contamination in the Seveso incident. Drug Metab Rev 13:407–422.

Fiedler H, Hutzinger O (1992) Sources and sinks of dioxins: Germany. Chemosphere 25:1487–1491.

Filonow AB, Jacobs LW, Mortland MM (1976) Fate of polybrominated biphenyls (PBBs) in soils. Retention of hexabromobiphenyl in for Michigan soils. J Agric Food Chem 24:1201–1204.

Firestone D, Rees J, Brown NL, Barron RP, Damico JN (1972) Determination of polychlorodibenzo-p-dioxins and related compounds in commercial chlorophenols. J Assoc Offic Anal Chem 55:85–92.

Fishbein L (1984) An overview of environmental and toxicological aspects of aromatic hydrocarbons. I. Benzene. Sci Total Environ 40:189–218.

Frank R, Fish N, Sirons GJ, Walker J, Orr HL, Leeson S (1982) Residues of polychlorinated phenols and anisoles in broilers raised on contaminated wood-shaving litter. Poult Sci 62:1559–1565.

Freeman RA, Schroy JM (1989) Comparison of the rate of TCDD transport at Times Beach and Eglin AFB. Chemosphere 18:1305–1312.

Fries GF, Marrow GS (1975a) Retention and excretion of 2,3,7,8-tetrachlorodibenzo-p-dioxin by rats. J Agric Food Chem 23:265–269.

Fries GF, Marrow GS (1975b) Excretion of polybrominated biphenyls into the milk of cows. J Dairy Sci 58:947–951.

Fries GF (1977) The kinetics of halogenated hydrocarbon retention and elimination in dairy cattle. In: Ivie GW, Dorough HW (eds) Fate of Pesticides in the Large Animal, Academic Press, New York, pp 159–173.

Fries GF, Marrow GS (1981) Chlorobiphenyl movement from soil to soybean plants. J Agric Food Chem 29:757–759.

Fries GF (1982) Potential polychlorinated biphenyl residues in animal products from application of contaminated sewage sludge to land. J Environ Qual 11:14–20.

Fries GF, Marrow GS, Snow PA (1982a) Soil ingestion by dairy cattle. J Dairy Sci 65:611–618.

Fries GF, Marrow GS, Snow PA (1982b) Soil ingestion by swine as a route of contaminant exposure. Environ Toxicol Chem 1:201–204.

Fries GF (1985a) Bioavailability of soil-borne poybrominated biphenyls ingested by farm animals. J Toxicol Environ Hlth 16:565–579.

Fries GF (1985b) The PBB episode in Michigan: An overall appraisal. CRC Crit Rev Toxicol 16:105–156.

Fries GF, Jacobs LW (1986) Evaluation of residual polybrominated biphenyl contamination present on Michigan farms in 1978. Research Report 477, Michigan State University, Agricultural Experiment Station, East Lansing, MI.

Fries GF (1987) Assessment of potential residues in foods derived from animals exposed to TCDD-contaminated soil. Chemosphere 16:2123–2128.

Fries GF, Paustenbach DJ (1990) Evaluation of potential transmission of 2,3,7,8-tetrachlorodibenzo-p-dioxin contaminated incinerator emissions to humans via foods. J Toxicol Environ Hlth 29:1–43.

Fries GF (1991) Organic contaminants in terrestrial food chains. In: Jones KC (ed) Organic Contaminants in the Environment. Elsevier, New York, pp 207–236.

Fries GF, Marrow GS (1992) Influence of soil properties on the uptake of hexachlorobiphenyls by rats. Chemosphere 24:109–113.

Fries GF (1994) The sensitivity of measurements of soil ingestion by livestock to assumptions concerning forage intake and digestibility. J Anim Sci 72(Suppl. 1): 275.

Furukawa K, Matsumura F (1979) Microbial metabolism of polychlorinated biphenyls. Studies on the relative degradability of polychlorinated biphenyl components by *Alkaligenes sp.* J Agric Food Chem 24:251–256.

Furukawa K, Tomizuka N, Kamibayashi A (1979) Effect of chlorine substitution on the bacterial metabolism of various polychlorinated chlorinated biphenyls. Appl Environ Microbiol 38:301–310.

Gaggi C, Bacci E (1985) Accumulation of chlorinated hydrocarbon vapors in pine needles. Chemosphere 14:451–456.

Gallo MA, Scheuplein RJ, van der Heijden RA (1991) Biological Basis for Risk Assessment of Dioxins and Related Compounds. Bandbury Report 35, Cold Spring Harbor Laboratory Press, Plainview, NY.

Gough TA, McPhail MF, Webb KS, Wood BJ, Coleman RF (1977) An examination of some foodstuffs for the presence of volatile nitrosamines. J Sci Food Agric 28:345–351.

Graham PR (1973) Phthalate platicizers — How and why they are used. Environ Hlth Perspect 3:13–15.

Haque R, Schmedding DW, Freed VH (1974) Aqueous solubility, adsorption, and vapor behavior of polychlorinated biphenyl Aroclor 1254. Environ Sci Technol 8:139–142.

Harrad SJ, Jones KC (1992) A source inventory and budget for chlorinated dioxins and furans in the United Kingdom environment. Sci Total Environ 126:89–107.

Harrison GW, Weber JB, Baird JV (1976) Herbicide phytoxicity as affected by selected properties of North Carolina soils. Weed Sci 24:120–126.

Healy WB, Ludwig TG (1965) Wear of sheep's teeth. I. The role of ingested soil. N Z J Agric Res 8:737–752.

Healy WB, Cutress TW, Michie C (1967) Wear of sheep's teeth. IV. Reduction of soil ingestion and tooth wear by supplementary feeding. N Z J Agric Res 10:201–209.

Healy WB (1968) Ingestion of soil by dairy cows. N Z J Agric Res 11:487–499.

Healy WB, Drew KR (1970) Ingestion of soil by hoggets grazing swedes. N Z J Agric Res 13:940–944.

Healy WB, McCabe WJ, Wilson GF (1970) Ingested soil as a source of microelements for grazing animals. N Z J Agric Res 13:503–521.

Helling CS, Isensee AR, Woolson EA, Ensor PDJ, Jones GE, Plimmer JR, Kearney PC (1973) Chlorodioxins in pesticides, soils, and plants. J Environ Qual 2:171–178.

Herrmann R, Baumgartner I (1987) Regional variation of selected polyaromatic and chlorinated hydrocarbons over the South Island of New Zealand, as indicated by their content in *Pinus radiata* needles. Environ Pollut 46:63–72.

Hitch RK, Day HR (1992) Unusual persistence of DDT in some western USA soils. Bull Environ Contam Toxicol 48:259–264.

Hughes BJ, Forsell JH, Sleight SD, Kuo C, Shull LR (1985) Assessment of pentachlorophenol-toxicity in newborn calves: Clinicopathology and tissue residues. J Anim Sci 61:1587–1063.

Hulster A, Marschner H (1993) Transfer of PCDD/PCDF from contaminated soils to food and fodder crop plants. Chemosphere 27:439–446.

Hutzinger O, Safe S, Zitko V (1972) Photochemical degradation of chlorobiphenyls (PCBs). Environ Hlth Perspect 1:15–20.

International Agency for Research on Cancer (1983) Polynuclear aromatic hydrocarbons. Part 1. Chemical, environmental and experimental data. Working group on the evaluation of carcinogenic risk of chemicals to humans, Lyon, France. Vol 32.

Jackson DR, Roulier MH, Grotta HM, Rust SW, Warner JS, Arthur MF, Deroos FL (1985) Leaching potential of TCDD in contaminated soils. U. S. Environmental Protection Agency, Report No. 600/9–85/013.

Jacobs LW, O'Connor GA, Overcash MA, Zabik MJ, Rygiewicz P, Machno P, Munger S, Elseewi, AA (1987) Effects of trace organics in sewage sludges on soil-plant systems and assessing their risk to trace elements to the food chain. In: Page AL, Logan TJ, Ryan JA (eds) Land Application of Sludge. Lewis Publishers, Inc., Chelsea, MI, pp 101–143.

Jensen DJ, Getzendaner ME, Hummel RA, Turley J (1983) Residue studies for (2,4,5-trichlorophenoxy)acetic acid and 2,3,7,8-tetrachlorodibenzo-p-dioxin in grass and rice. J Agric Food Chem 31:118–122.

Jones D, Safe S, Morcom E, Holcomb M, Coppock C, Ivie W (1989) Bioavailability of grain and soil-borne tritiated 2,3,7,8-tetrachlorodibenzo-p-dioxin (TCDD) administered to lactating Holstein cows. Chemosphere 18:1257–1263.

Jones KC, Grimmer G, Jacob J, Johnston AE (1989a) Changes in the polynuclear aromatic hydrocarbon content of wheat grain and pasture grassland over the last century from on site in the U.K. Sci Total Environ 78:117–130.

Jones KC, Stratford JA, Waterhouse KS, Furlong ET, Giger W, Hites RA, Schaffner C, Johnston AE (1989b) Increases in the polynuclear aromatic hydrocarbon content of an agricultural soil over the last century. Environ Sci Technol 23:95–101.

Jones KC, Stratford JA, Waterhouse KS, Vogt NB (1989c) Organic contaminants in Welsh soils: Polynuclear aromatic hydrocarbons. Environ Sci Technol 23:540–550.

Jones KC, Sanders G, Wild SR, Burnett V, Johnston AE (1992) Evidence for a decline of PCBs and PAHs in rural vegetation and air in the United Kingdom. Nature 356:137–140.

Jones SG, Brown KW, Deuel LE, Donnelly KC (1979) Influence of simulated rainfall on the retention of sludge heavy metals by the leaves of forage crops. J Environ Qual 8:69–72.

Jury WA, Spencer WF, Farmer WJ (1983) Behavior assessment model for trace organics in soil: Model description. J Environ Qual 12:558–564.

Kapila S, Yanders AF, Orazio CE, Meadows JE, Cerlesi S, Clevenger TE (1989) Field and laboratory studies on the movement and fate of tetrachlorodibenzo-p-dioxin in soil. Chemosphere 18:1297–1304.

Kearney PC, Woolson EA, Ellington CP (1972) Persistence and metabolism of chlorodioxins in soils. Environ Sci Technol 6:1017–1019.

Kearney PC, Amundson ME, Beynon KI, Drescher N, Marco GJ, Miyamoto J, Murphy JR, Oliver JE (1980a) Nitrosamines and Pesticides. A special report on the occurrence of nitrosamines as terminal residues resulting from agriculture of certain pesticides. Pure Appl Chem 52:449–526.

Kearney PC, Oliver JE, Konston A, Fiddler W, Pensabene JW (1980b) Plant uptake of dinitroanaline herbicide related nitrosamines. J Agric Food Chem 28:633–635.

Kenaga EE (1980) Correlation of bioconcentration factors of chemicals in aquatic and terrestrial organisms with their physical and chemical properties. Environ Sci Technol 14:553–556.

Kimbrough RD, Falk H, Stehr P, Fries G (1984) Health implications of 2,3,7,8-tetrachlorodibenzo-p-dioxin (TCDD) contamination of residential soil. J Toxicol Environ Hlth 14:47–93.

Kinzell JH, McKenzie RM, Olson BA, Kirsch DG, Shull LR (1985) Metabolic fate of (U-^{14}C) pentachlorophenol in a lactating dairy cow. J Agric Food Chem 33: 827–833.

Kluwe WM (1982) Overview of phthalate ester pharmacokinetics in mammalian species. Environ Hlth Perspect 45:3–10.

Koester CJ, Hites RA (1992) Wet and dry deposition of dioxins and furans. Environ Sci Technol 26:1375–1382.

Kofoed AD, Nemming O, Brunfeldt, Nebelin E, Thomsen J (1981) Investigations on the occurrence of nitrosamines in some agricultural and horticultural products. Acta Agric Scand 31:40–48.

Laflamme RE, Hites RA (1978) The global distribution of polycyclic aromatic hydrocarbons in recent sediments. Geochim Cosmochim Acta 42:289–303.

Langer HG, Brady TP, Briggs PR (1973) Formation of dibenzodioxins and other condensation products from chlorinated phenols and derivatives. Environ Hlth Perspect 5:3–8.

Ligon WV, Dorn SB, May RJ, Allison MJ (1989) Chlorodibenzofuran and chlorodibenzo-p-dioxin levels in Chilean mummies dated to about 2800 years before the present. Environ Sci Technol 23:1286–1290.

Lusky K, Stoyke M, Henke G (1992) Untersuchungen zum Vorkommen von polyzyklischen, aromatischen Kohlenwasserstoffen (PAH) im Futter und bei landwirtschaftlichen Nutztieren. Arch Lebensmittelhyg 43:67–68.

Martin WE (1964) Losses of Sr^{90}, Sr^{89} and I^{131} from fallout of contaminated plants. Radiat Bot 4:275–281.

Mayland HF, Florence AR, Rosenau RC, Lazar VA, Turner HA (1975) Soil ingestion by cattle on semiarid range as reflected by titanium analysis of feces. J Range Mgt 28:448–452.

McCrady JK, MacFarlane C, Gander LK (1990) The transport and fate of 2,3,7,8-TCCD in soybean and corn. Chemosphere 21:359–376.

McCrady JK, Maggard SP (1993) Uptake and photodegradation of 2,3,7,8-tetrachlorodibenzo-p-dioxin sorbed to grass forage. Environ Sci Technol 27:343–350.

McFarland VA, Clarke JU (1989) Environmental occurrence, abundance, and potential toxicity of polychlorinated biphenyl congeners: Considerations for a congener specific analysis. Environ Hlth Perspect 81:225–239.

McGrath D, Poole DBR, Fleming GA, Sinnott J (1982) Soil ingestion by grazing sheep. Irish J Agric Res 21:135–145.

McLachlan MS, Thoma H, Reissinger M, Hutzinger O (1989) PCDD/F in an agricultural food chain. Part 1: PCDD/F mass balance in a lactating cow. Chemosphere 20:1013–1020.

McLachlan MS (1993) Exposure toxicity equivalents (ETEs): A plea for more environmental chemistry in dioxin risk assessment. Chemosphere 27:483–490.

McLachlan MS (1993) Mass balance of polychlorinated biphenyls and other organochlorine compounds in a lactating cow. J Agric Food Chem 41:474–480.

Miller GC, Herbert VR, Miille MJ, Mitzel R, Zepp RG (1989) Photolysis of octchlorodibenzo-*p*-dioxin on soils: Production of 2,3,7,8-TCDD. Chemosphere 18:1265–1274.

Mirvish SS (1970) Kinetics of dimethylamine nitrosation in relation to nitrosamine carcinogenesis. J Natl Cancer Inst 44:633–639.

Modica R, Fiume M, Guaitani A, Bartosek I (1983) Comparative kinetics of benz(a)anthracene, chrysene and triphenylene in rats after oral administration. I. Study with single compounds. Toxicol Lett 18:103–109.

Moghissi AA, Marland RE, Congel FJ, Eckerman KF (1980) Methodology for environmental human exposure and health risk assessment. In: Haque R (ed) Dynamics, Exposure and Hazard Assessment of Toxic Chemicals. Ann Arbor Science Publishers, Ann Arbor, MI, pp 477–489.

Mohler K, El-Refai SM (1981) Das Nitrosaminproblem in der menschlichen Umwelt: Bestimmung von fluchtigen N-Nitrosaminen in Maissilage und Milch mit der Nitramin-Methode Vergleich mit der Thermoluminiscenz-Analyse. Z Lebensm Unters Forsch 172:449–453.

Mumma RO, Raupach DR, Waldman JP, Hotchkiss JH, Gutenmann WH, Bache CA, Lisk DJ (1983) Analytical survey of elements and other constituents in central New York State sewage sludges. Arch Environ Contam Toxicol 12:581–587.

Mumma RO, Raupach DR, Waldman JP, Tong SSC, Jacobs ML, Babish JG, Hotchkiss JH, Wszolek PC, Gutenmann WH, Bache CA, Lisk DJ (1984) National survey of elements and other constituents in municipal sewage sludges. Arch Environ Contam Toxicol 13:75–83.

Nash RG, Woolson EA (1967) Persistences of chlorinated hydrocarbon insecticides in soils. Science 157:924–927.

Nash RG, Woolson EA (1968) Distribution of chlorinated insecticides in cultivated soil. Soil Sci Soc Am Proc 32:525–527.

Nash RG, Beall ML (1980) Distribution of silvex, 2,4-D, and TCDD applied to turf in chambers and field plots. J Agric Food Chem 28:614–623.

Nisbet ICT, Sarofim AF (1972) Rates and routes of transport of PCBs in the environment. Environ Health Perspect 1:21–38.

Noble A (1990) The relation between organochlorine residues in animal feeds and residues in tissues, milk and eggs: A review. Aust J Exp Agric 30:145–154.

O'Connor GA, Chaney RL, Ryan JA (1991) Bioavailability to plants of sludgeborne toxic organics. Rev Environ Contam Toxicol 121:129–155.

Olie K, Vermeulen PL, Hutzinger O (1977) Chlorodibenzo-*p*-dioxins and chlorodibenzofurans are trace components of fly ash and flue gas of some municipal incinerators in the Netherlands. Chemosphere 6:455–459.

Olie K, van den Berg M, Hutzinger O (1983) Formation and fate of PCDD and PCDF from combustion processes. Chemosphere 12:627–636.

Olling M, Derks HJGM, Berender PLM, Liem AKD, de Jong APJM (1991) Toxicokinetics of eight ^{13}C-labelled polychlorinated-*p*-dioxins and -furans in lactating cows. Chemosphere 23:1377–1385.

Osweiler GD, Olesen B, Rottinghaus GE (1983) Plasma pentachlorophenol concentrations in calves exposed to treated wood in the environment. Am J Vet Res 45:244–246.

Pahren, HR (1980) Overview of the problem. In: Bitton G, Damron BL, Edds GT, Davidson JM (eds) Sludge Health Risks of Land Application. Ann Arbor Science, Ann Arbor, MI, pp 1-6.

Paltineanu IC, Apostol I (1974) Possibilities of using the neutron probe method for water application efficiency studies in sprinkler and furrow irrigation. In: Isotope and Radiation Techniques in Soil Physics and Irrigation Studies. International Atomic Energy Agency, Vienna, pp 477-506.

Park KS, Sims RC, Dupont RR, Doucette WJ, Matthews JE (1990) Fate of PAH compounds in two soil types: Influence of volatilization, abiotic loss and biological activity. Environ Toxicol Chem 9:187-195.

Parkin TB, Codling EE (1990) Rainfall distribution under a corn canopy: Implications for managing agrochemicals. Agron J 82:1166-1169.

Parmar D, Srivastava SP, Srivastava SP, Seth PK (1985) Hepatic mixed function oxidases and cytochrome P-450 contents in rat pups exposed to di-(2-ethylhexyl)phthalate through mother's milk. Drug Metab Dispos 13:368-370.

Paterson S, Mackay D, Bacci E, Calamari D (1991) Correlation of the equilibrium and kinetics of leaf-air exchange of hydrophobic organic chemicals. Environ Sci Technol 25:866-871.

Paustenbach DJ, Wenning RJ, Lau V, Harrington NW, Rennix DK, Parsons AH (1992) Recent developments on the hazards posed by 2,3,7,8-tetrachlorodibenzo-p-dioxin in soil: Implications for setting risk-based cleanup levels at residential and industrial sites. J Toxicol Environ Hlth 36:103-149.

Peterson JH (1991) Survey of di-(2-ethylhexyl)phthalate plasticizer contamination of retail Danish milks. Food Addit Contam 8:701-706.

Petreas MX, Goldman LR, Hayward DG, Chang RR, Flattery JJ, Wiesmuller T, Stephens RD, Fry DM, Rappe C, Bergek S, Hjelt M (1991) Biotransfer and bioaccumulation of PCDD/PCDFs from soil: Controlled feeding studies of chickens. Chemosphere 23:1731-1741.

Pocchiari F, Di Domenico A, Silvano V, Zapponi G (1983) Environmental impact of the accidental release of tetrachlorodibenzo-p-dioxin (TCDD) at Seveso (Italy). In: Coulston F, Pocchiari F (eds) Accidental Exposure to Dioxins. Academic Press, New York, pp 5-38.

Poiger H, Schlatter C (1980) Influence of solvents and adsorbents on dermal and intestinal absorption of TCDD. Food Cosmet Toxicol 18:477-481.

Rappe C, Kjeller, LO, Anderson R (1989) Analyses of PCDDs and PCDFs in sludge and water samples. Chemosphere 19:13-20.

Rappe C (1992) Sources of PCDDs and PCDFs. Introduction. Reactions, levels, patterns, profiles and trends. Chemosphere 25:41-44.

Reischl A, Reissinger M, Thoma H, Hutzinger O (1989) Accumulation of organic constituents by plant surfaces: Part IV. Plant surfaces: A sampling system for atmospheric polychlorodibenzo-p-dioxins (PCDD) and polychlorodibenzofurans (PCDF). Chemosphere 18:561-568.

Riordan C (1983) A decade of progress. In: Page, AL, Gleason TL, Smith JE, Iskander IK, Sommers LE (eds) Utilization of Municipal Wastewater and Sludge on Land. University of California, Riverside, CA, pp 15-21.

Robens J, Anthony HD (1980) Polychlorinated biphenyl contamination of feeder cattle. J Am Vet Med Assoc 177:613-615.

Rose JQ, Ramsey JC, Wentzler TH, Hummel RA, Gehring PJ (1976) The fate of 2,3,7,8-tetrachlorodibenzo-*p*-dioxin following single and repeated oral doses to the rat. Toxicol Appl Pharmacol 36:209–226.

Russell RS (1963) The extent and consequences of the uptake and by plants of radioactive nuclides. Ann Rev Plant Physiol 14:271–294.

Ruzo LO, Sundstrom G, Hutzinger O, Safe S (1976) Photodegradation of polybrominated biphenyls (PBB). J Agric Food Chem 24:1062–1065.

Ryan JA, Bell RM, Davidson JM, O'Connor GA (1988) Plant uptake of non-ionic chemicals from soils. Chemosphere 17:2299–2323.

Ryan JJ, Lizotte R, Sakuma T, Mori B (1985) Chlorinated dibenzo-*p*-dioxins, chlorinated dibenzofurans, and pentachlorophenol in Canadian chicken and pork samples. J Agric Food Chem 33:1021–1026.

Shiu WY, Mackay D (1986) A critical review of aqueous solubilities, vapor pressures, Henry's law constants, and octanol-water partition coefficients of the polychlorinated biphenyls. J Phys Chem Ref Data 15:911–929.

Shiu WY, Doucette W, Gobas FAPC, Andren A, Mackay D (1988) Physical-chemical properties of chlorinated dibenzo-*p*-dioxins. Environ Sci Tchnol 22: 651–658.

Shull LR, Foss M, Anderson CR, Feighner K (1981) Usage patterns of chemically treated wood on Michigan dairy farms. Bull Environ Contam Toxicol 26:561–566.

Smith RM, O'Keefe PW, Aldous KM, Briggs R, Hilker DR, Conner S (1992) Measurements of PCDFs and PCDDs in air samples and lake sediments at several locations in upstate New York. Chemosphere 25:95–98.

Steiner JL, Kanemasu ET, Clark RN (1983) Spray losses and partitioning of water under a center pivot system. Trans Am Soc Agric Eng 26:1128–1134.

Stevens JB, Gerbec EN (1988) Dioxin in the agricultural food chain. Risk Analysis 8:329–335.

Strachan WMJ, Eriksson G, Kylin H, Jensen S (1994) Oganochlorine compounds in pine needles: Methods and trends. Environ Toxicol Chem 13:443–451.

Strek HJ, Weber JB, Shea PJ, Mrozek E, Overcash MR (1981) Reduction of polychlorinated biphenyl toxicity and uptake of carbon-14 activity by plants through the use of activated carbon. J Agric Food Chem 29:288–293.

Subcommittee on Feed Intake (1987) Predicting Feed Intake of Food-Producing Animals. National Academy Press, Washington, DC.

Suntio LR, Shiu WY, Mackay D, Seiber JN, Glotfelty D (1988) Critical review of Henry's law constants for pesticides. Rev Environ Contam Toxicol 103:1–59.

Tashiro C, Clement RE, Stocke BJ, Radke L, Cofer WR, Ward P (1990) Preliminary report: Dioxins and furans in prescribed burns. Chemosphere 20:1533–1566.

Tate RL, Alexander M (1976) Resistence of nitrosamines to microbial attack. J Environ Qual 5:131–133.

Terplan G, Hallermayer E, Kalbfus W, Unsinn P, Gartner CD, Heerdegen C (1978) Untersuchungen zum Vorkommen von Nitrosaminen in Futtermitteln, Milch und Milchprodukten. Milchwissenschaft 33:142–145.

Terplan G, Bucsis L, Heerdegen CH (1980) Nitrosamine in Futter, Milch, und Milchprodukten. Arch Lebensmittelhyg 31:1–5.

Thomas JA, Northrup SJ (1982) Toxicity and metabolism of monoethylhexyl phthalate and diethylhexyl phalate: A survey of recent literature. J Environ Hlth Toxicol 9:141–152.

Thornton I, Abrahams P (1981) Soil ingestion — A major pathway of heavy metals into livestock grazing contaminated land. Sci Total Environ 28:287–294.

Travis CC, Arms AD (1988) Bioconcentration of organics in beef, milk, and vegetation. Environ Sci Technol 22:271–274.

Travis CC, Hattemer-Fry HA (1988) Uptake of organics by aerial parts: A call for research. Chemosphere 17:277–283.

Tuinstra LGM, Vreman K, Roos AH, Keukens HJ (1981) Excretion of certain chlorobiphenyls into the milk fat after oral administration. Neth Milk Dairy J 35:147–157.

Tuinstra LGM, Roos AH, Berende PLM, van Rhijn JA, Traag WA, Mengelers MJB (1992) Excretion of polychlorinated dibenzo-*p*-dioxins and furans in milk of cows fed on dioxins in the dry period. J Agric Food Chem 40:1772–1776.

U.S. Department of Agriculture (1992) Agricultural Statistics. U.S. Government Printing Office, Washington, DC.

U.S. Environmental Protection Agency (U.S. EPA) (1989) Exposure Factors Handbook. EPA-600/8-89/043, Office of Health and Environmental Assessment, Washington, DC.

U.S. EPA (1990) Assessment of Risks from Exposure of Humans, Terrestrial and Avian Wildlife, and Aquatic Life to Dioxins and Furans from Disposal and Use of Sludges from Bleached Kraft Sulfite Pulp and Paper Mills. EPA-560/5-90/013, Office of Toxic Substances and Office of Solid Waste, Washington, DC.

Wakeham SG, Schaffner C, Giger W (1980) Polycyclic aromatic hydrocarbons in recent lake sediments-I. Compounds having anthropogenic origins. Geochim Cosmochim Acta 44:403–413.

Weber JB (1972) Interaction of organic pesticides with particulate matter in aquatic and soil systems. Adv Chem Series 111:55–120.

Weerasinghe NCA, Gross ML, Lisk DJ (1985) Polychlorinated dibenzodioxins and polychlorinated dibenzofurans in sewage sludge. Chemosphere 14:557–564.

Wickstrom K, Tolonen K (1987) The history of airborne polycyclic aromatic hydrocarbons (PAH) and perylene as recorded in dated lake sediments. Water Air Soil Pollut 32:155–175.

Wild SR, McGrath SP, Jones KC (1990a) The polynuclear aromatic hydrocarbon (PAH) content of archived sewage sludges. Chemosphere 20:703–716.

Wild SR, Waterhouse KS, McGrath SP, Jones KC (1990b) Contaminants in an agricultural soil with a known history of sewage sludge amendments: Polynuclear aromatic hydrocarbons. Environ Sci Technol 24:1706–1711.

Wild SR, Obbard JP, Munn CI, Berrow ML, Jones KC (1991) The long-term persistence of polynuclear hydrocarbons (PAHs) in an agricultural soil amended with metal-contaminated sewage sludge. Sci Total Environ 101:235–253.

Wild SR, Jones KC (1993) Biological and abiotic losses of polynuclear aromatic hydrocarbons (PAHs) from soils freshly amended with sewage sludge. Environ Toxicol Chem 12:5–12.

Willett LB, Hess JF (1975) Polychlorinated biphenyl residues in silos in the United States. Residue Reviews 55:135–147.

Willett LB, Durst HI (1978) Effects of PBBs on cattle. IV. Distribution and clearance of components of FireMaster BP-6. Environ Hlth Perspect 23:67–74.

Willett LB, Liu TY, Fries GF (1990) Reevaluation of polychlorinated biphenyl concentrations in milk and body fat of lactating cows. J Dairy Sci 73:2136–2142.

Willett LB, O'Donnell AF, Durst HI, Kurz MM (1993) Mechanisms of movement of organochlorine pesticides from soils to cows via forages. J Dairy Sci 76:1635–1644.

Witherspoon JP, Taylor FG (1970) Interception and retention of simulated fallout by agricultural plants. Hlth Phys 19:493–499.

Withey JR, Law FCP, Endrenyi L (1991) Pharmacokinetics and bioavailability of pyrene in the rat. J Toxicol Environ Hlth 32:429–447.

Zedeck MS (1980) Polycyclic aromatic hydrocarbons: A review. J Environ Pathol Toxicol 3:537–567.

Manuscript received May 2, 1994; accepted June 6, 1994.

Pesticide Residues in Olive Oil

Chaido Lentza-Rizos* and Elizabeth J. Avramides*

Contents

I. Introduction

The olive tree (*Olea europea*) is a subtropical tree that grows in temperate climates and may be cultivated in arid and infertile areas. According to the International Olive Oil Council's (IOOC) estimate, 750 million trees are grown worldwide, covering a total area of 9 million hectares (Kyritsakis 1988). Of these, 98% are grown in the countries surrounding the Mediterranean Sea, whereas the remaining 2% are grown in America. The olive tree is cultivated for its fruit, which is rich in oil. The total world production of olive fruits is estimated to be 8–9 million metric tons, of which a small part (0.4–0.7 million metric tons) is used for the preparation of table olives (fruits eaten after various processing treatments). The remainder is processed to provide olive oil. Olive oil is the oil obtained solely from the fruit of the olive tree and is classified, according to the IOOC (1992), as follows:

(1) Virgin olive oil is oil obtained from the fruit of the olive tree solely by mechanical or other physical means under conditions, particularly thermal conditions, that do not lead to alteration of the oil. It does not undergo any treatment other than washing, decantation, centrifugation, or filtration. Virgin olive oil that is fit for consumption "as is" and may be referred to as "natural," includes:

---

*Pesticide Residue Laboratory, Benaki Phytopathological Institute and National Agricultural Research Foundation, 145 61 Kiphissia, Athens, Greece.

© 1995 by Springer-Verlag New York, Inc.
Reviews of Environmental Contamination and Toxicology, Vol. 141.

Extra virgin olive oil is virgin oil that has an organoleptic rating of 6.5 or more and a free acidity, expressed as oleic acid, of not more than 1 g/100 g, with due regard for the other criteria established in this standard.

Fine virgin olive oil is virgin oil that has an organoleptic rating of 5.5 or more and a free acidity, expressed as oleic acid, of not more than 1.5 g/100 g, with due regard for the other criteria in this standard.

Semifine virgin olive oil (or ordinary virgin olive oil) is virgin oil that has an organoleptic rating of 3.5 or more and a free acidity, expressed as oleic acid, of not more than 3.5 g/100 g, with due regard for the other criteria in this standard.

(2) Refined olive oil is oil obtained from virgin oil by refining methods that do not lead to alteration of the initial glyceridic structure.

(3) Olive oil is oil consisting of a blend of refined oil and virgin oil fit for consumption as is.

Virgin olive oil retains all of its important characteristics of taste and flavor, is of high biological and nutritional value compared with other fats and oils, and is readily digested. It has been used over the years as an important nutritional food item, as a medication, and as a cosmetic. Its consumption rate is high in the producing countries, ranging from 20 kg/person/annum in Greece to 2.2 kg/person/annum in Turkey (Kyritsakis 1988). World production of olive oil during the years 1981–1986 ranged from 1.4 to 2.2 million metric tons per annum, of which 95% was produced in the Mediterranean countries of Spain, Italy, Greece, Portugal, Tunisia, Algeria, and Morocco, in decreasing order of importance.

Olive trees are attacked by several pests and diseases. The key insect pests of Mediterranean olives are the olive fruit fly *Bactocera* (*Dacus*) *oleae*, the olive-kernel borer or olive moth, *Prays oleae*, and black scale, *Saissetia oleae*. Although *B. oleae* is considered the most serious pest, all three are widely distributed in the region and occur on olives at population levels causing important economic damage. Of the less important insect pests, some occur in particular areas or under specific conditions at population levels that cause serious damage, e.g., *Euphyllura olivina*, *Zeuzera pyrina*, *Aspidiotus nerii*, and *Resseliella oleisuga*. Others, although occurring only occasionally at such levels, cause serious problems by disrupting the biological balance of the ecosystem, e.g., *Parlatoria oleae*, *Leucapsis riccae*, and *Philippia follicularis* (Katsoyanos 1992).

The most important diseases are fungal diseases caused by the fungi *Verticillium dahliae*, *Spilocaea oleagina* (*Cycloconium oleaginum*), *Leveillula taurica*, *Cercospora cladosporioides*, *Gleosporium olivarum*, and *Camarosporium dalmatica* (*Macrophoma dalmatica*) (Elena 1990). In addition, the bacterium *Pseudomonas syringae* pv. *Savastanoi*, which is responsible for olive knot disease, causes considerable losses. Furthermore, many weeds compete with olive trees for nutrients and water or impede harvest. Therefore, the control of pests, diseases, and weeds is necessary to ersure satisfactory production in terms of both quantity and quality and is generally carried out with chemical pesticides. Some growth regulators also are authorized for use in some countries, with the aim of increasing fruit setting by decreasing the fall of blossom and fruit or to facilitate mechanical harvest by enhancing fruit abscission.

Given the toxicological properties of currently used pesticides and the high consumption rate of olive products in many countries, especially those surrounding the Mediterranean Sea, attention has been drawn to the problem of pesticide residues in olive fruits and olive oil. Olive fruits undergo severe treatment to become edible, and this is very likely to eliminate or greatly reduce residues, but the same is not true for virgin olive oil.

A review of data on insecticide residues in olives and olive oil was published by Alessandini (1962). At that time, many organochlorine insecticides, which are known to concentrate in fatty substrates, were used. Since then, treatment patterns have changed, further scientific work has been performed, and new methods of analysis have been developed. This review summarizes the available literature through 1993.

II. Methods of Analysis
A. Extraction and Cleanup

Multiresidue methods suitable for fatty substrates are generally used for the determination of residues in olive oil. These methods are based on partitioning between hexane or light petroleum and acetonitrile, and gas–liquid chromatography (GLC) determination [Association of Official Analysis of Chemicals (AOAC) 1984; U.S. Food and Drug Administration (FDA) 1982; Health and Welfare Canada 1990]. Other methods, such as the sweep codistillation technique (Renvall and Akerblom 1971), semipreparative liquid chromatography (LC) silica gel cleanup, and the Unitrex system also have been used. Other extraction techniques may also be applied, such as the alumina blending technique (Gillespie and Walters 1984) and gel-permeation chromatography (Hopper 1982; Blaha and Jackson 1985). In 1988, the AOAC committee on residues, established to study multiresidue methods and to make proposals for official methods, undertook an investigation of alternatives to acetonitrile partitioning.

Initial results for semipreparative LC silica gel cleanup showed poor separation of some organophosphorus pesticides from oils, while attempts to use a reverse-phase separation with a C-18 column and an acetonitrile mobile phase resulted in recovery problems with organochlorine pesticides. Initial attempts to use the Unitrex system for vegetable oils were unsuccessful, possibly because the oils were oxidized or degraded (Sawyer 1988). Subsequent studies with four refined vegetable oils, including olive oil, showed excellent cleanup and approximately 80–90% recoveries for six of eight pesticides tested, i.e., three benzene hexachloride (BHC) isomers, chlorpyrifos, dieldrin, and heptachlor epoxide (see Appendix for chemical names).

Endrin and p, p'-DDT recoveries were variable, probably because of their susceptibility to degradation under certain chromatographic conditions (Sawyer 1988). Guvener et al. (1978) extracted phosphamidon residues through partitioning between *n*-hexane and water, followed by a second partitioning of the aqueous phase with chloroform and cleanup on a florisil column.

The hexane-acetonitrile extraction system, followed by column cleanup on florisil, was also used by Albi and Navas (1985a) for the determination of residues of the following 15 organophosphorus insecticides: diazinon, dimethoate and its metabolite dimethoxon, fenthion, phosmet, fitios, malathion, parathion, methyl parathion, methidathion, ethion, trichlorfon, fenitrothion, monocrotophos, and carbophenothion.

Ferreira and Tainha (1983) used the AOAC method for the extraction of the pesticides diazinon, dimethoate, methidathion, parathion, and phosphamidon from 2 g analytical samples, along with a further cleanup step on two florisil columns (deactivated with 2% and 1% water), and obtained recoveries in the range 73–100%.

Di Muccio et al. (1987) employed ready-to-use, disposable minicolumns of Kieselghur-type material (Extrelut®) for the direct extraction and cleanup of the following nine organophosphorus insecticides: diazinon, etrimfos, methyl chlorpyrifos, methyl pirimiphos, chlorpyrifos, bromophos, ethyl bromophos, malathion, and fenitrothion from fatty extracts, including olive oil. Up to 2 g oil samples could be accommodated, with the amount of oil released into the eluate being of the order of 1.4–2.4%. However, Morchio et al. (1992) were not able to confirm this finding, and much higher quantities of oil were coextracted. The connection of a ready-to-use silica-gel (Sep-Pak®) cartridge and a C-18 column to the Extrelut column improved the cleanup efficiency of the system, with only 0.14% of oil coextracted from a 1.8 g oil sample (Di Muccio et al. 1990).

One weak point in all of these methods is that they accommodate only a small sample size of 1–3 g. Further cleanup steps are required for the larger samples needed to achieve good GLC sensitivity, because coextracted oil interferes with the analysis. This is especially true for capillary columns, which deteriorate rapidly if oil is allowed to pass through them. Leone et al. (1990) succeeded in processing 5 g analytical olive oil samples containing the pesticides dimethoate, phosphamidon, formothion, fenthion, and the pyrethroid deltamethrin using successive partitionings between hexane and aqueous (5%) acetonitrile and cleanup by solid phase extraction (Superclean, LC-Florisil).

The extraction method developed by Lentza-Rizos and Avramides (1990) handles 10 g analytical samples and achieves satisfactory cleanup of the extract by liquid–liquid partitioning alone, making it possible to avoid time-consuming column cleanup. This is achieved by the addition of a small amount of water (1%) to the hexane–acetonitrile solvent system and the inclusion of a further hexane backwash of the acenonitrile extract. In this way, the amount of oil coextracted was limited to 0.3–1%.

This method, as initially developed, included an oxidation step, which allowed the determination of all the oxidative metabolites of fenthion, the most important insecticide used in olive groves in Greece. However, analysis of a large number of olive oil samples has shown that the total fenthion residue is given to a good approximation by the sum of the parent and sulfoxide metabolite residues alone and that the oxidation step is therefore not usually necessary (Lentza-Rizos and Avramides 1991). This extraction and cleanup method has been used for the insecticides fenthion and its metabolite fenthion sulfoxide, dimethoate, diazinon, ethyl azinphos, chlorpyrifos, methidathion, parathion, and methyl parathion (Lentza-Rizos and Avramides 1991). Recently, it has also been successfully used for formothion, etrimfos, and phosphamidon residues and for the herbicides atrazine and simazine (Lentza-Rizos, unpublished data).

Morchio and De Andreis (1989) adapted a technique that allows oil samples to be injected directly into a gas chromatograph equipped with a capillary column. A glass precolumn with an oil recovery tank in the injector port prevents column contamination. The concept (including the design of the liner) was first introduced for packed columns by Morchio (1982) and used for the determination of hexane residues in oils. It has subsequently been used for the determination of antioxidants and chlorinated solvents in addition to organophosphorus insecticides. A 1 μL sample, previously diluted 1 : 1 with acetone, is directly injected through the split-splitless system, using the injection technique of solvent flash. According to the authors, contamination of the GC column is avoided because oil is retained in the special precolumn tank, which can easily be changed, and recoveries and limits of detection are satisfactory. This appears to be a simple, fast, and cost-effective method, which drastically reduces the time and expense of analysis, but it is not described in detail in the paper.

Some additional information has been given in a subsequent paper (Morchio et al. 1992) in which it is also pointed out that particular care is required during injection of the oil/acetone mixture into the precolumn in order to achieve vaporization of the pesticides and deposition of the oil on the walls of the precolumn. The precolumn should be changed at the end of each working day. To avoid eventual transfer of volatile oil components from the injector into the GC column, which would decrease its resolution efficiency, it is recommended that the initial part of the 25 m column be joined using a press-fit to a 2–3 m piece of GC column, with no stationary phase. This causes microdrops of oil to be collected preferentially in the press-fit and not to enter the main column. When the column resolution deteriorates, i.e., the peaks become wider and the retention time longer, the initial piece of column may be changed and the press-fit washed and reused.

This method was tested on an analytical standard of dimethoate, cis- and trans-phosphamidon, diazinon, methyl parathion, malathion, fenthion, and methidathion dissolved in refined olive oil. The recoveries were higher than those obtained in the same laboratory with the most common conventional methods. A limit of determination of 0.002–0.003 mg/kg is stated for a new FPD-50 Carlo Erba detector, but it is not specified to which compounds this applies. Further information about this technique is given by Grob et al. (1993), who investigated the critical parameters of the method and proposed modifications for its optimization.

Mestres et al. (1978) developed a specific method for the determination of the pyrethroid deltamethrin, according to which 10 mL of oil is extracted with petroleum ether saturated with dimethyl sulfoxide (DMSO). The extract is partitioned successively with DMSO saturated with petroleum ether, and cleanup of the DMSO extract is effected by partitioning with an aqueous solution of NaCl and ethyl acetate, followed by adsorption chromatography on florisil or alumina.

B. Determination

GLC with phosphorus specific detectors (nitrogen phosphorus detector (NPD), thermionic specific detector (TSD), alkali flame ionization detector (AFID), flame photometric detector (FPD) P-mode) is the method of choice for the qualitative and quantitative determination of residues of organophosphorus insecticides. Most researchers have used borosilicate-glass columns of 2–4 mm i.d. and varying length, packed with 4% SE-30 (Renvall and Akerblom 1971), 10% DC-200 (Guvener et al. 1978), 1.5% OV-17 + 1.95% QF-1 (Ferreira and Tainha 1983), 2% DC-200 + 3.7% QF-1 (Ferreira and Tainha 1990), 5% OV-17 (Lentza-Rizos and Avramides 1990, 1991), or 5% QF-1 (Di Muccio et al. 1987, 1990). Capillary columns of fused silica have been used: SPB1 of 0.25 mm i.d. (Leone et al. 1990) and SPB 608 of 0.32 mm i.d. (Morchio and De Andreis 1989), with film thickness 0.25 μm and length 30 m, and PS 255 of 0.32 mm i.d. with film thickness 0.4 μm and length 25 m (Morchio et al. 1992). For deltamethrin, according to Mestres et al. (1978), various stationary phases may be used (DC 200, OV 1 or OV 101), but separation of the *cis*- and *trans*-isomers is achieved only on DC LSX 30295 with an electron capture detector.

III. Insecticide Residue Trials

Most of the research work on pesticide residues in olive oil involves insecticides because of their toxicity and the necessity for repeated applications. Some treatments are carried out late in autumn, i.e., near harvest, which makes it more likely that detectable residues will remain in the oil. Insecticides are applied to olive groves either as full coverage treatments (runoff) or as bait sprays.

A. Full Coverage Sprays

Trials in Portugal with the organophosphorus insecticides diazinon, dimethoate, methidathion, parathion, and phosphamidon have ascertained that residue concentrations in olive oil are related to the number of treatments, the preharvest interval, and the fat-solubility of the compound. Water-soluble pesticides (dimethoate, phosphamidon) pass into the aqueous phase during the extraction of oil from olives and are therefore discarded. However, fat-soluble pesticides (parathion, methidathion, and diazinon) were found to concentrate in the oil with a concentration factor of 3–5 (Ferreira and Tainha 1983, 1990). As shown in Table 1, residues in the oil 41 d (1980) and 42 d (1982) postapplication at the recommended application rate were of the order of 0.70 and 0.99 (parathion), 1.40 and 2.00 (methidathion), and 4.90 and 2.70 (diazinon) mg/kg, respectively. Residues of dimethoate and phosphamidon were not detectable even 1 d postapplication.

Leone et al. (1990) reached the same conclusion regarding the importance of the fat- or water-solubility of pesticides. Olive trees were treated once with the organophosphorus insecticides dimethoate, phosphamidon, formothion, and fenthion and with the pyrethroid deltamethrin at the recommended and twice the recommended application rates. The oil was produced either in the laboratory or by industrial extraction in an oil mill. The laboratory preparation was carried out at several time intervals after application using a mechanical procedure reproducing the conditions of an industrial olive oil press. The industrial extraction was undertaken 53 d postapplication. As shown in Table 2, the residues of fat-soluble fenthion were always much higher than those of water-soluble insecticides applied at the same rate.

Gambacorta et al. (1993) applied dimethoate at two application rates (19 and 28 g a.i./hL water) on plots of 20 trees. Two treatments were made 37 d apart, in late August and early October (1990). Olive samples of 500 g were collected 1, 7, 17, and 38 d after the second application and processed into oil in the laboratory. At 52 d postapplication, the whole plots were harvested and the oil was produced in an industrial oil mill. No detectable residues (< 0.005 mg/kg) were found from 38 d postapplication (Table 3).

In 1993, dimethoate was applied by the owner of a Greek olive grove as a full coverage (runoff) treatment. Four treatments were made at 14-d intervals, starting in August, at the application rate of 50 g a.i./hL water. One more treatment was made in November, at 160 g a.i./hL water. The crop was harvested in mid-December and the oil extracted immediately. Dimethoate residues in the oil produced were 0.10 mg/kg (Lentza-Rizos, unpublished data).

B. Bait Sprays

The bait spray technique is extensively used in some countries. A mixture of a food attractant and an organophosphorus insecticide formulation is applied from the ground or air. In Greece, bait spraying is carried out by government organizations, with a frequency determined by the results of monitoring the olive fly population by mass trapping. Hydrolyzed proteins

Table 1. Organophosphorus insecticide residues in olives and oil after full-coverage treatments in Portugal (after Ferreira and Tainha 1983, 1990).

Pesticide	Year	Application rate (g a.i./hL)	Dates of treatments (day/month/year)	Preharvest interval (d)	Residues (mg/kg) Olives	Oil
Parathion	1980	36	7/10/1980	14	0.37	1.10
				21	0.46	1.30
				28	0.45	1.30
				35	0.29	0.80
				41	0.29	0.70
	1982	36	6/10/1982	14	0.59	4.20
				21	0.48	2.90
				28	0.35	1.60
				35	0.24	1.40
				42	0.21	0.99
				49	N.A.	1.00
Methidathion	1980	60	7/10/1980	14	0.81	4.30
				21	0.51	2.00
				28	0.54	1.80
				35	0.30	1.60
				41	0.16	1.40
	1982	60	6/10/1982	14	2.20	7.60
				21	1.80	5.80
				28	0.96	4.20
				35	0.64	2.30
				42	0.50	2.00
				49	0.55	2.10
				55	0.52	1.70

Pesticide	Year		Date	Day		
Diazinon	1980	90	2/9/1980 and 7/10/1980	14	1.50	6.60
				21	2.60	8.40
				28	2.30	9.60
				35	1.80	6.40
				41	0.91	4.90
	1982	90	6/10/1982	14	1.20	5.50
				21	1.10	3.70
				28	0.84	3.60
				35	0.62	3.10
				42	0.85	2.70
				49	0.47	2.00
				55	0.56	1.70
Dimethoate	1980	60	2/9/1980 and 7/10/1980	1	5.30	N.D.
				7	3.10	N.D.
				14	1.50	N.D.
				21	0.78	N.D.
				28	0.41	N.A.
				35	0.28	N.A.
				41	0.30	N.A.
Phosphamidon	1980	60	2/9/1980 and 7/10/1980	1	7.20	N.D.
				7	3.40	N.D.
				14	2.10	N.D.
				21	0.39	N.D.
				28	0.25	N.A.
				35	N.D.**	N.A.

N.A. = Not analyzed.
N.D. = Not detectable.

Table 2. Insecticide residues in olive oil after full-coverage sprays in Italy (after Leone et al. 1990).

Pesticide	Date of treatment (day/month/year)	Application rate (g a.i./hL)	Preharvest interval (d)	Residues in oil (mg/kg)	
				Laboratory pressing	Industrial pressing
Dimethoate	16/11/1988	60	8	0.48	
			20	0.33	
			32	0.19	
			47	0.09	
			53		0.06
Phosphamidon	16/11/1988	60	8	0.24	
			20	0.07	
			32	<0.01	
			47	<0.01	
			53		<0.01
Phormothion	16/11/1988	60	8	0.15	
			20	0.04	
			32	<0.01	
			47	<0.01	
			53		<0.01
Deltamethrin	16/11/1988	57	8	0.23	
			20	0.12	
			32	0.04	
			47	<0.01	
			53		<0.01
Fenthion	16/11/1988	60	8	1.72	
			20	1.48	
			32	0.99	
			47	0.56	
			53		0.43

Table 3. Dimethoate residues in oil after full-coverage spray (after Gambacorta et al. 1993).

Application rate	Time (d)				
(g a.i./hL)	1	7	17	38	52
28	1.50	0.33	0.02	N.D.	N.D.
19	0.80	0.19	0.01	N.D.	N.D.

N.D. = Not detectable (< 0.005 mg/kg).

are used as attractants, with dimethoate or fenthion formulations. From the ground, the product is applied by knapsack sprayers on a part of one of every two trees in each orchard. The spray mixture consists of 2–4% w/w attractant plus 0.3% w/w active ingredient (a.i.) insecticide. Approximately 300 mL of this solution is applied to a small part of the canopy of the treated trees. In this way, 30 L spray mixture/ha are applied, i.e., 90 g a.i./ ha [low volume (LV) application]. From the air, the mixture is applied by airplane or helicopter at 10 L/ha, consisting of 4–12% w/w attractant plus 0.9% w/w insecticide, again giving 90 g a.i./ha [very low volume (VLV) application]. Approximately 85 million trees are treated annually. The method used (air or ground) depends on the availability of workers and the suitability of the location for each method. For ecological reasons, a strong preference has developed recently for ground treatment. 100–120 tons a.i. of fenthion and 80–90 tons a.i. of dimethoate are used annually.

Fenthion.   Trials using LV or VLV applications were carried out in Greece on the islands of Samos and Crete to study the possibility of shortening the statutory preharvest interval of 30 d. The results are summarized in Table 4. In Samos, the experimental orchards consisted of a local, medium-sized variety of olives suitable both for table consumption and for oil extraction. Those in Crete were of an extensively grown, small-sized variety destined only for oil extraction. Two formulations of fenthion were tested, emulsifiable concentrate (EC) and water miscible concentrate (EW). The concentration of attractant was 3.5% w/w for LV and 4% for VLV applications. Each trial had three replicates and each replicate had three plots of approximately 10,000 trees.

In Samos, six treatments were made in 1984, starting in July and ending in October. Samples weighing 4 kg were collected from each plot 12 d after the last treatment, and the oil was extracted in the laboratory of the Institute of Subtropical Plants and the Olive Tree in Crete using a simulated industrial technique. The residues of fenthion (parent compound) in fresh olives were 0.97 mg/kg after the application of LV sprays with EC formulation and 0.70 mg/kg after the application of LV sprays with EW formulation. Those in the oil were approximately three times higher.

Table 4. Residues of fenthion (parent compound) in olives and oil after bait spraying (90 g a.i./ha) (Lentza-Rizos, unpublished data).

Location, sample size	Variety	Application mode and dates of treatments (day/month/year)	Interval between last treatment and harvest (d)	Residues mg/kg			
				Olives		Oil	
				EC	EW	EC	EW
Samos 4 kg	Medium-sized for table consumption and oil production	Ground (LV)[a] 15/7/1984 11/8/1984 2/9/1984 21/9/1984 8/10/1984 26/10/1984	12	0.97 ±0.15	0.70 ±0.14	2.71[c] ±0.70	1.86[c] ±0.47
Crete 100 kg	Small-sized for oil production	Ground (LV)[a] 18/6/1986 20/7/1986 8/8/1986 17/9/1986 24/10/1986	21	—	—	0.21[d] ±0.02	0.18[d] ±0.04
Crete 4 kg	Small-sized for oil production	Air VLV[b] 27/7/1984 24/8/1984 5/9/1984 13/9/1984 3/10/1984 17/11/1984	12 24	— —	0.18 ±0.08 0.10 ±0.03	— —	0.57[c] ±0.27 0.31[c] ±0.09
Crete 100 kg		Air VLV[b] 18/7/1986 8/8/1986 13/11/1986	21	—	—	0.63[d] ±0.12	0.62[d] ±0.13

[a]Knapsack sprayer 0.3% w/w a.i., 3.5% w/w attractant, 30:1 mixture/ha (300 mL/tree).
[b]0.9% w/w a.i., 4% w/w attractant, 10:1 mixture/ha.
[c]Oil extracted in the laboratory.
[d]Oil extracted in an industrial olive mill.

In 1986, five LV applications of insecticide were made in Crete, starting in June and ending in October. Residues of fenthion (parent compound) in the oil were 0.21 mg/kg for EC and 0.18 mg/kg for EW formulation, 21 d postapplication.

VLV treatments by airplane were made on the island of Crete in 1984 and 1986. In 1984, six treatments were made with EW formulation starting at the end of July and continuing until mid-November. Both fresh olives and the oil produced from them were analyzed. Residues in olives after the application of VLV sprays with EW formulation were 0.18 mg/kg at 12 d postapplication and 0.10 mg/kg at 24 d postapplication. The same concentration factor of 3 was observed for the oil. In 1986, three treatments were made on each plot with EC or EW formulation. Fenthion residues in the oil at 21 d postapplication were the same for the two formulations (0.62 mg/kg) (Lentza-Rizos, unpublished data). During the 1984 trials, the difficulty of obtaining fruits from the upper parts of the trees led to their underrepresentation in the samples collected. This problem was resolved for the 1986 trials, for which 100 kg samples were picked from 35 trees in each plot. The large size of each sample made it possible to carry out the processing in an industrial oil mill to obtain the oil.

Cabras et al. (1993) studied the persistence and fate of fenthion in olives and olive products. In oil produced from olives treated three or five times with LV bait sprays of 800 g a.i./hL, and collected 54 d after the last application, residues of the parent compound were almost three times higher than those in the olives (1.01 and 0.34 mg/kg, respectively, for the plot treated three times and 2.29 and 0.72 mg/kg, respectively, for the plot treated five times). The parent compound was the most important residue. From the five oxidative metabolites analyzed, fenthion oxon sulfoxide and fenthion oxon sulfone were absent. Fenthion sulfoxide was present at relatively high concentration both in oil and olives (0.25 and 0.19 mg/kg, respectively, for the plot treated three times, and 0.27 and 0.51 mg/kg, respectively, for the plot treated five times). Fenthion sulfone and fenthion oxon were present at very low concentrations in oil and olives.

Dimethoate.   Albi and Navas (1985b) analyzed samples of olives and olive oil extracted in the laboratory after aerial application of dimethoate on an olive grove of 50 ha. The application rate was 1000 g a.i./hL water and hydrolyzed protein at 2.5% was added to the spray mixture, applied at 20 L/ha (200 g a.i./ha). Olive samples of 4 kg were collected 2 hr (day 0) and 2, 5, 7, 9, 12, and 15 d after treatment and processed into oil. A subsample of the day 0 sample was analyzed for dimethoate residues in the olives. The results (Table 5) showed that at day 0 the oil contained only 22% of the dimethoate present in the olives from which the oil was produced and that, 1 wk after treatment, the mean concentration in oil was very low (0.008 mg/kg).

Table 5. Dimethoate residues (mg/kg) after aerial bait application (0.2 kg a.i./ha) (after Albi and Navas 1985b).

Time (d)	Olives	Oil
0	0.715	0.166
2	N.A.	0.090
5	N.A.	0.042
7	N.A.	0.008
9	N.A.	N.A.
12	N.A.	N.A.
15	N.A.	N.A.

N.A. = Not analyzed.

Phosphamidon. In Turkey in 1976, olive trees were treated twice (September and November) with phosphamidon ULV 100 EC at 500 g a.i./ha. Olive fruit samples collected 7, 14, and 21 d after the second application contained 0.05 mg/kg, 0.035 mg/kg, and nondetectable residues, respectively, whereas the oil produced at harvest (presumably at preharvest interval 21 d) had no detectable residues (limit of detection not stated) (Guvener et al. 1976). Details of the application technique are not given. However, the ultra low volume (ULV) bait method uses 1200–1500 mL/ha spray mixture with a pesticide and a food attractant in the proportion w/w of 1 : 13.3 (without water).

IV. Fungicide Residue Trials

Copper fungicides. Only one report of fungicide residue trials appears in the literature. It refers to treatments to control olive leaf spot (*Spilocaea oleagina*) using Bordeaux mixture or "fixed" copper fungicides (cupric hydroxide, tribasic copper sulfate, copper oxide, or copper ammonium carbonate) to compare the efficacy of fungicides and determine suitable application rates. Copper residue analyses were made of olive leaves, and these results are not relevant to residues in olive oil and consumer protection (Teviotdale et al. 1989).

V. Herbicide Residue Trials

Glyphosate. An experimental application of glyphosate at 300 g a.i./ha was made in Spain under the olive tree canopy in late autumn when olive fruits were on the ground, a nonauthorized use. The oil from the olives contained nondetectable (<0.05 mg/kg) glyphosate residues (Valera and Costa 1990).

VI. Monitoring Data

Sweden. Between 1965 and 1967, 50 samples of olive oil imported from France, Italy, Spain, and the United Kingdom were obtained from Swedish shops, and analyzed for diazinon, dimethoate, malathion, parathion, and methyl parathion residues. Only diazinon was found in 8 (16%) of the samples, ranging from 0.01 to 0.13 mg/kg. The limit of detection for the other pesticides was 0.02 mg/kg (Renvall and Akerblom 1971).

Greece. During 1984–1985, 45 samples were analyzed for fenthion (parent compound only) and dimethoate residues. Most of them (37) were virgin oil and the remaining were refined. The latter did not contain detectable residues [limit of determination (LOD) 0.01 mg/kg for fenthion and 0.05 mg/kg for dimethoate]. Fenthion was detected in 23 samples of virgin oil (62%), ranging from 0.01 to 0.30 mg/kg (Lentza-Rizos 1985). Further research in 1988–1990 involved collection and analysis of virgin oil samples from all over Greece (560 during 1988–1989 and 70 during 1989–1990). The analytical method used permitted the determination of total fenthion residues (parent compound plus its oxidative metabolites), dimethoate, methidathion, methyl parathion, and ethyl azinphos (Lentza-Rizos and Avramides 1990). No dimethoate residues were detected in any of the samples (LOD 0.01 mg/kg), presumably because of its water solubility. For the first and second years, 50% and 21%, respectively, of the samples contained no detectable total fenthion residues, whereas 4% and 6%, respectively, had residues exceeding the FAO/WHO Codex Alimentarius Maximum Residue Limit (MRL) (1 mg/kg) (FAO/WHO 1994).

Treatment records of the samples with a fenthion concentration exceeding the MRL showed that high residues resulted from bait spraying from the ground or full-coverage (runoff) treatments by the producers. In the former case, the amount of spray mixture applied and the proportion of trees treated appear to be in violation of the recommendation that only one of each two trees should be treated on only a small part of its foliage. The mean concentration for the first year was 0.216 mg/kg and for the second year 0.256 mg/kg. Fenthion (parent compound) was the most important residue in fresh oil samples, whereas aged samples contained a higher amount of the most important metabolite, fenthion sulfoxide. The concentration of the oxygen analogs (P=O metabolites) of fenthion was <5% of the total residue concentration in most cases.

Of the other insecticides determined, ethyl azinphos, methidathion, and methyl parathion were detected in a few samples. Of these, only ethyl azinphos appeared to cause a problem as a result of its use in a few areas in late autumn, in violation of the official recommendation that it be used on olive trees only until the end of August.

From the mean total fenthion concentration of the 2-yr study (0.236 mg/kg oil) and estimated olive oil consumption per person per year (20 kg), a daily dietary exposure of 0.0002 mg/kg body weight was calculated for a mean body weight of 60 kg. This is lower than the acceptable daily intake (ADI) for man given by FAO/WHO (0.001 mg/kg/d) (FAO 1981). Olive oil is the only likely source of fenthion because olives are essentially the only edible crop on which this insecticide is used.

Follow-up research carried out in 1991–1992 on 30 samples of commercially packed oil and 115 virgin oil samples collected from individual growers involved analysis for ethyl azinphos, chlorpyrifos, diazinon, dimethoate, fenthion, fenthion sulfoxide, methidathion, ethyl parathion, and methyl parathion residues. Commercially packed oil samples contained either no detectable residues or low concentrations of fenthion, fenthion sulfoxide, and chlorpyrifos. More than one-half of the virgin oil samples collected from individual growers contained no detectable residues. The others contained mostly fenthion and its sulfoxide metabolite (Lentza-Rizos 1994).

Italy. A total of 154 samples of extra virgin oil and 36 of "olive oil" (a mixture of 80% refined and 20% extra virgin oil) produced locally or imported from Greece or Spain (year not stated) were analyzed for organophosphorus insecticide residues (diazinon, dimethoate, phosphamidon, methyl parathion, malathion, ethyl parathion, fenthion, and methidathion). Fenthion and methidathion were found to be the most widespread in Italian and Greek extra virgin oils, presumably due to their fat solubility and persistence. Spanish samples had negligible residues (< 30 ppb), which was attributed to the fact that, in Spain, treatments are basically made with water-soluble dimethoate. The olive oil samples contained only 10–20% of the concentrations detected in extra virgin oil samples. This was attributed to elimination of residues during the refining process (Morchio and De Andreis 1989).

In a continuation of the monitoring program, 174 samples of extra virgin oil were analyzed with the latest multiresidue method developed by these authors. The conclusions were similar to those of the previous study, i.e., that fenthion and methidathion were the most common residues detected (79% and 57%, respectively, of the total), with mean concentrations 0.073 mg/kg and 0.040 mg/kg, respectively. There were negligible residues of dimethoate, diazinon, phosphamidon, methyl parathion, and malathion, except in two cases where diazinon residues at 0.20 and 0.29 mg/kg were determined (Morchio et al. 1992).

VII. Fate of Residues

Storage. After preliminary studies (Lentza-Rizos 1987) had shown that the concentration of fenthion (parent compound) in stored olive oil decreased with time, a follow-up study of its decay was carried out (Lentza-

Rizos et al. 1994). Concentrations of both the parent compound and its sulfoxide metabolite were monitored because of the toxicological importance of fenthion's oxidative metabolites. It was found that the concentration of the parent compound decreased slowly with time, following a double-phase, first-order decay model both in the freezer and the laboratory, but that the total concentration (fenthion plus sulfoxide) remained unchanged over a year of storage. These results preclude the use of storage as a possible decontamination technique for olive oil containing higher than toxicologically acceptable residues of fenthion and emphasize the importance of determining fenthion sulfoxide as well as fenthion in analysis of oil.

Refining.   Vioque et al. (1973) submitted olive oil spiked with HCH, lindane, heptachlor, epoxyheptachlor, aldrin, dieldrin, endrin, and p,p'-DDT at the 1 mg/kg level for refining. Significant amounts of these pesticides were eliminated during the refining process, and the most important step from this point of view is deodorization. At a temperature of approximately 240 °C, most of these pesticides are completely eliminated from the oil during deodorization. Part of the p,p'-DDT is transformed during deodorization to p,p'-DDD, which is eliminated, whereas 20–25% of the initial p,p'-DDT is detected in the refined oil.

The fate of organophosphorus pesticide residues during refining has only recently been studied, and results indicate that residue elimination takes place here as well. Morchio and De Andreis (1989) found that the "coupe" samples they analyzed contained much lower concentrations of organophosphorus insecticides than virgin oil, which the authors attributed to the effect of refining. A similar conclusion was reached by Lentza-Rizos (1994), who analyzed the deodorization distillate of an olive oil refinery. This distillate, of mass 3 tons was collected over a period of 1 yr after the treatment of 5000 tons. The analysis revealed a high (138 mg/kg) fenthion sulfoxide concentration.

Morchio et al. (1992) investigated the fate of organophosphorus insecticide residues during the classic alkali refining process and the physical neutralization process of raw virgin oil. They spiked 200 kg of oil (3.2% acidity) with dimethoate, diazinon, *cis*- and *trans*-phosphamidon, methyl parathion, malathion, fenthion, and methidathion and submitted both to the classical refining process in a pilot plant and to the physical neutralization process. During the neutralization phase of the traditional alkali refining process, dimethoate and phosphamidon were totally degraded, methidathion was reduced by 28%, and 50% of the residues of the other organophosphorus compounds were eliminated. During decolorization, the percentage reduction varied from 95% (diazinon) to 30% (methidathion). During the deodorization phase, degradation was almost complete for all compounds (approximately 5% of the initial concentration was detected in the refined oil).

Using the physical refining process, phosphamidon was completely eliminated and dimethoate and diazinon were reduced by 90%, whereas the
other compounds were removed to a lesser extent than with the alkali process. However, the deodorization step performed using different methods
was shown to be a very effective procedure for eliminating organophosphorus insecticide residues.

VIII. Maximum Residue Levels

In order to protect consumers, many countries have established maximum
residue limits (MRLs) (also called maximum residue levels or tolerances) in
agricultural products destined for human consumption. These national limits are established on the basis of appropriate supervised trial data and take
into account the long-term toxicity of each compound, as expressed by its
ADI for man. MRLs are fixed only for crops for which an authorized/
registered or officially recommended treatment exists in the MRL-setting
country. Thus, pesticides or pesticide/crop combinations not registered in a
particular country have no tolerance level, which means zero tolerance. In
such cases, an import tolerance may be established, but this requires sufficient residue and toxicological data.

This system of national MRLs very often leads to obstacles in the international trade of agricultural commodities. For that reason, the United
Nations organizations FAO and WHO have undertaken an international
harmonization within the framework of Codex Alimentarius. However, the
proposed Codex MRLs are not always accepted by all member states, which
often find that either these limits are not based on sufficient residue data or
that the need for many pesticide uses is not well established or is overesti-
mated. This is to some extent true, especially in the case of olive oil.

The fact that olive trees are grown mainly in countries with limited
resources means that work on pesticide residues relevant to the establishment of a MRL is limited. Furthermore, these countries have for years
applied a somewhat flexible and undemanding system of pesticide registration, with the result that pesticide-producing companies have not submitted
pesticide residue data sufficient for the establishment of a tolerance in olive
oil. Some residue data have been reported in FAO Plant Production and
Protection Papers (yearly evaluations of the Joint Meetings of the FAO
Panel of Experts on Pesticide Residues in Food and the Environment and
the WHO Expert Group on Pesticide Residues). These data are not reviewed here because they are in most cases highly summarized. However,
they have enabled the establishment by FAO/WHO Codex Alimentarius
Commission of MRLs in olive oil for the five insecticides given in Table 6
(FAO/WHO 1994).

In the U.S., under section 408 of the Federal Food, Drug, and Cosmetic
Act (FFDCA), the EPA establishes tolerances, or exemptions from tolerances when appropriate, for pesticide residues in raw agricultural commodi-

Table 6. Codex Alimentarius MRLs for olives and olive oil.

	MRLs (mg/kg)				
	Olives			Oil	
Pesticide	Fresh	Processed	Not specified	Virgin	Refined
Carbaryl	10	1			
Deltamethrin			0.1		
Diazinon			2[a]	2[a]	
Dimethoate	1	0.05			0.05[b]
Fenthion and metabolites			1	1	
Methidathion			2	2	
Paraquat			1		
Parathion			0.5	2	
Permethrin (sum of isomers)			1		
Pirimiphos, methyl			5		

[a]Withdrawal recommended, 1993.
[b]Limit of determination.

ties. Maximum permissible levels of pesticide residues in processed foods are established under section 409. However, section 409 tolerances are required only for certain pesticide residues in processed food. Under section 402 of the FFDCA, no section 409 tolerance is required if any pesticide residue in a processed food resulting from use of a pesticide on a raw agricultural commodity is below the tolerance for that pesticide in or on the raw agricultural commodity (flow-through provision). Thus, a section 409 tolerance is only necessary to prevent foods from being deemed adulterated when the concentration of the pesticide residue in a processed food is greater than the tolerance prescribed for the raw agricultural commodity (U.S. EPA 1993). Any pesticide for which no tolerance has been established has a zero tolerance level. For olive oil, the flow-through provision has been applied to date by the U.S. EPA, and tolerances are set only for the raw agricultural commodity (olive fruits). However, this policy does not take into account the concentration of many fat-soluble pesticides in olive oil, and it may ultimately have serious implications for trade of this commodity.

Summary

The attacks of pests and diseases and the presence of weeds make it necessary to apply pesticides to olive trees to ensure crop protection. Residues of these compounds may remain and contaminate the oil produced. For the analysis of pesticide residues in olive oil, the most common methods are multiresidue methods for fatty substrates, based on partitioning between hexane or light petroleum and acetonitrile. Recently, other methods have been applied, such as ready-to-use, disposable minicolumns or direct injection of oil into a capillary gas chromatograph equipped with a precolumn with an oil recovery tank.

Although several pesticides are registered in oil-producing countries for use on olive trees, available literature on the level and fate of residues is very limited. However, it is clear that fat-soluble pesticides tend to concentrate in the oil, both after full coverage and bait spraying, and their use close to harvest should therefore be avoided. Because it is sometimes necessary to use such pesticides late in autumn because of their effectiveness in cases of severe attack, residue trials should be carried out to determine the residue concentration in oil and to set a reasonable preharvest safety interval.

Data produced by such trials would permit the establishment of MRLs (tolerances) in olive oil to cover cases where the residues, although relatively high, are not of toxicological significance for consumers (risk assessment). Such is the case with corn oil and the fat-soluble insecticide methyl pirimiphos, registered in the U.S. for use on corn. The U.S.EPA tolerance for methyl pirimiphos in corn is 8 mg/kg, whereas it is 11 times higher (88 mg/kg) for corn oil because it is known to concentrate in the oil. Similar provisions for olive oil, based on data from residue trials according to Good Agricultural Practice, the long-term toxicity of each pesticide as expressed by its ADI for man, and olive oil consumption patterns, would facilitate international trade of this commodity. On the other hand, because of the high dietary and health value of olive oil, it is desirable that toxic pesticide residues be kept as low as possible. Therefore, it would be preferable not to rely only on chemical pest control treatments but to develop and apply alternative plant protection techniques such as Integrated Pest Management (IPM). To that aim, several international organizations, i.e., European Economic Community (EEC), International Olive Oil Council (IOOC), and the International Organization for Biological Control (IOBC) have initiated coordinated research activities, with FAO and the United Nations Development Program as the major contributors. According to FAO, considerable knowledge of IPM for olive groves has already been acquired from research activities in various places, mostly in the EEC Mediterranean countries. By making adjustments to the methods developed to suit local requirements and conditions, olive IPM schemes may be immediately implemented (Katsoyannos 1992).

References

Albi T, Navas MA (1985a) Phosphorus insecticide determination in oils. Grasas y aceites 36:48–54 (in Spanish).

Albi T, Navas MA (1985b) Insecticide residues in edible fats. III. Analysis of phosphorus insecticides. Grasas y aceites 36:373–375 (in Spanish).

Alessandrini ME (1962) Insecticide residues in olive oils and table olives from efforts to control the olive fly. Residue Reviews 1:92–111.

Association of Official Analytical Chemists (AOAC) (1984) Official Methods of Analysis, Assoc Offic Anal Chem, 14th Ed. AOAC, Arlington, VA, p 533.

Blaha JJ, Jackson PJ (1985) Multiresidue method for quantitative determination of organophosphorus pesticides in foods. J Assoc Offic Anal Chem 68:1095–1099.

Cabras P, Garau VL, Melis M, Pirisi FM, Spanedda L (1993) Persistence and fate of fenthion in olives and olive products. J Agric Food Chem 41:2431–2433.

Di Muccio A, Cicero AM, Camoni I, Pontecorvo D, Dommarco R (1987) On-column partition cleanup of fatty extracts for organophosphate pesticide residue determination. J Assoc Offic Anal Chem 70:106–108.

Di Muccio A, Ausili A, Vergori L, Camoni I, Dommarco R, Gambetti L, Santilio A, Vergori F (1990) Single-step multicartridge cleanup for organophosphate pesticide residue determination in vegetable oil extracts by gas chromatography. Analyst 115:1167–1169.

Elena K (1990) Fungal diseases of olive tree in Greece. Technical Bull. No 11. Benaki Phytopathological Institute, Kiphissia, Greece (in Greek).

Ferreira JR, Tainha AM (1983) Organophosphorous insecticide residues in olives and olive oil. Pestic Sci 14:167–172.

Ferreira JR, Tainha AM (1990) Residues of the fat-soluble insecticides diazinon, methidathion, and parathion in olives and olive oil. Poster 08D-19. Seventh International Congress of Pesticide Chemistry, IUPAC, Hamburg, August 5–10.

FAO (1981) Plant production and protection, Paper 26, Suppl Pesticide Residues in Food — 1980, Rome, p 230.

FAO/WHO Codex Alimentarius Commission (1994) Codex Committee on Pesticide Residues. Status of Codex Maximum Residue Limits for pesticides in food and animal food: CX/PR 2-1994.

Food and Drug Administration (FDA) (1982) General methods for fatty foods. Pesticide Analytical Manual, Volume I, section 231. U.S. Dept. of Health and Human Services, FDA, Washington, DC.

Gambacorta G, Pizza M, La Notte E (1993) The use of dimethoate in Dacus oleae (Gmel) control: Residue problem in olive oil. Riv Ital Sostanza Grasse 70:289–294 (in Italian).

Gillespie AM, Walters SM (1984) Alumina blending technique for separation of pesticides from lipids. J Assoc Offic Anal Chem 67:290–294.

Grob K, Biedermann M, Guiffré AM (1993) Determination of organophosphorus insecticides in edible oils and fats by splitless injection of the oil into GC (injector-internal headspace analysis). Z Lebensm Unters Forsch (in press).

Guvener A, Kortimur G, Turker O, Kucukkalipsi F (1978) Phosphamidon residue investigation in olives after treatment with ULV bait spray technique by airplane against Dacus oleae Gmel. Plant Prot Res Ann, 12:63–64 (in Turkish).

Health and Welfare Canada, Health Protection Branch (1990) National Pesticide Residue Limits in Foods. Bureau of Chemical Safety, Food Directorate, Ottawa, Ontario.

Hopper ML (1982) Automated gel permeation system for rapid separation of industrial chemicals and organophosphate and chlorinated pesticides from fats. J Agric Food Chem 30:1038–1041.

International Olive Oil Council (1992) International trade standards applying to olive oil and olive-pomace oils. Resolution No. RES-3/66-IV/92.

Katsoyannos P (1992) Olive pests and their control in the Near East. FAO Plant Production and Protection, Paper 115. FAO, Rome.

Kyritsakis A (1988) Olive Oil. Agricultural Cooperatives Editions, Thessaloniki, Greece.

Lentza-Rizos Ch (1985) Organophosphorus insecticide residues in olive oil samples. Proceedings of the first Hellenic Entomological Congress, pp 173–177 (in Greek).

Lentza-Rizos Ch (1987) Preliminary observations on the dissipation of fenthion residues in olive oil. Proceedings of second Hellenic Entomological Congress, pp 141–149 (in Greek).

Lentza-Rizos Ch, Avramides EJ (1990) Determination of residues of fenthion and its oxidative metabolites in olive oil. Analyst 115:1037–1040.

Lentza-Rizos Ch, Avramides EJ (1991) Organophosphorus insecticide residues in virgin Greek olive oil, 1988–1990. Pestic Sci 32:161–171.

Lentza-Rizos CH, Avramides EJ, Roberts RA (1994) Persistence of fenthion residues in olive oil. Pestic Sci 40:63–69.

Lentza-Rizos Ch (1994) Monitoring pesticide residues in olive products. Part I: Insecticide residues in olives and oil. J Assoc Offic Anal Chem Int 77:1096–1100.

Leone AM, Liuzzi VA, Gambacorta G, La Notte E, Santoro M, Alviti F, Laccone G, Guario A (1990) Research on some organophosphate and pyrethroid insecticide residues in oil extracted from olives subjected to guided field control trials against Dacus oleae. Riv Ital Sostanze Grasse 67:17–28 (in Italian).

Mestres R, Chevallier Ch, Espinosa Cl, Cornet R (1978) Dosage des residus de decamethrine dans les produits vegetaux. Trav Soc Pharm Montpellier 38:183–192.

Morchio G (1982) Rapid GC determination of residual hexane in crude B olive oil and refined olive oil and B olive oil. Riv Ital Sostanze Grasse 59:335–340 (in Italian).

Morchio G, De Andreis R (1989) Rapid determination by capillary GC of organophosphorus residues in vegetable oils with special reference to olive oil. Actes du Congres International (Chevreul) pour l'Etude des Corps Gras, 2:608.

Morchio G, De Andreis R, Verga GR (1992) Investigation of organophosphorous residues in vegetable oils with special reference to olive oil. Riv Ital Sostanze Grasse 69:147–157 (in Italian).

Renvall S, Akerblom M (1971) Determination of organophosphorus pesticide residues in olive oil. FAO Plant Prot Bull 3:57–61.

Sawyer LD (1988) Multiresidue methods (Interlaboratory studies). J Assoc Offic Anal Chem 71:93–94.

Teviotdale BL, Sibbett GS, Harper DH (1989) Control of olive leaf spot by copper fungicides. Appl Agric Res 4:185–189.

United States Environmental Protection Agency (USEPA) (1993) Federal Register 58(23):7470.

Valera A, Costa J (1990) Safety of Sting SE in olive groves. Herbicide treatments when olives are lying on the ground. Actes de la Reunion de la Sociedad Espanola de Malerbologia, 1990:225–230 (in Spanish).

Vioque A, Albi T, Nosti M (1973) Residues of pesticides in edible fats. II. Elimination of chlorinated insecticides during refining process. Grasas y aceites 24:20–26 (in Spanish).

Manuscript received July 9, 1994; accepted July 14, 1994.

Appendix: Common and Chemical Names of Pesticides

aldrin: (1α, 4α, $4a\beta$, 5α, 8α, $8a\beta$)-1,2,3,4,10,10-hexachloro-1,4,4a,5,8,8a-hexahydro-1,4:5,8-dimethanonaphthalene (9CI). CAS No. [309-00-2].

atrazine: 6-chloro-N-ethyl-N'-(1-methylethyl)-1,3,5-triazine-2,4-diamine (9CI). CAS No. [1912-24-9].

azinphos, ethyl: O,O-diethyl S-[(4-oxo-1,2,3-benzotriazine-3(4H)-yl)methyl] phosphorodithioate (9CI). CAS NO. [2642-71-9].

BHC: 1,2,3,4,5,6-hexachlorocyclohexane (mixed isomers) (8&9CI). CAS No. [39284-22-5].

bromophos: O-(4-bromo-2,5-dichlorophenyl) O,O-dimethyl phosphorothioate (8&9CI). CAS No. [2104-96-3].

bromophos, ethyl: O-(4-bromo-2,5-dichlorophenyl) O,O-diethyl phosphorothioate (8&9CI). CAS No. [4824-78-6].

bromophos, methyl: see bromophos

carbaryl: 1-naphthalenyl methylcarbamate (9CI); (I)(8CI). CAS No [63-25-2].

carbophenothion: S-{[(4-chlorophenyl)thio] methyl} O,O-diethyl phosphorodithioate (9CI). CAS No. [786-19-6].

chlorpyrifos: O,O-diethyl O-(3,5,6-trichloro-2-pyridinyl) phosphorothioate (9CI). CAS No. [2921-88-2].

chlorpyrifos methyl: O,O-dimethyl O-(3,5,6-trichloro-2-pyridinyl) phosphorothioate (9CI). CAS No. [5598-13-0].

deltamethrin: {1R-[1α(S*),3α]}-cyano(3-phenoxyphenyl) methyl 3-(2,2-dibromoethenyl)-2,2-dimethylcyclopropanecarboxylate (9CI). CAS No. [52918-63-5].

diazinon: O,O-diethyl O-[6-methyl-2-(1-methylethyl)-4-pyrimidinyl] phosphorothioate (9CI). CAS No. [333-41-5].

dieldrin: (1aα,2β,2aα,3β,6β,6aα,7β,7aα)-3,4,5,6,9,9-hexachloro-1a,2,2a,3,6,6a,7,7a-octahydro-2,7:3,6-dimethanonaphth[2,3-b] oxirene (9CI). CAS No. [60-57-1].

dimethoate: O,O-dimethyl S-[2-(methylamino)-2-oxoethyl] phosphorodithioate (9CI). CAS No. [60-51-5].

endrin: (1aα,2β,2aβ,3α,6α,6aβ,7β,7aα)-3,4,5,6,9,9-hexachloro-1a,2,2a,3,6,6a,7,7a-octahydro-2,7:3,6-dimethanonaphth[2,3-b]oxirene (9CI). CAS No. [72-20-8].

ethion: S,S'-methylene bis(O,O-diethyl phosphorodithioate) (9CI). CAS No. [563-12-2].

etrimfos: O-(6-ethoxy-2-ethyl-4-pyrimidinyl)O,O-dimethyl phosphorothioate (9CI). CAS No. [38260-54-7].

fenitrothion: O,O-dimethyl O-(3-methyl-4-nitrophenyl) phosphorothioate (9CI). CAS No. [122-14-5].

fenthion: O,O-dimethyl O-[3-methyl-4-(methylthio)phenyl] phosphorothioate (9CI). CAS No. [55-38-9].

fitios: S-[2-(ethylamino)-2-oxoethyl] O,O-dimethyl phosphorodithioate (9CI). CAS No. [116-01-8].

formothion: S-[2-(formylmethylamino)-2-oxoethyl] *O,O*-dimethyl phosphorodithioate (9CI). CAS No. [2540-82-1].

glyphosate: *N*-(phosphonomethyl) glysine (8&9CI). CAS No. 1071-83-6].

heptachlor: 1,4,5,6,7,8,8-heptachloro-3a,4,7,7a-tetrahydro-4,7-methano-1H-indene (9CI). CAS No. [71-44-8].

HCH: see BHC.

lindane: ($1\alpha,2\alpha,3\beta,4\alpha,5\alpha,6\beta$)-gamma HCH. CAS No. [58-89-9].

malathion: diethyl[(dimethoxyphosphinothioyl)thio] butanedioate (9CI). CAS No. [121-75-5].

methidathion: S-{[5-methoxy-2-oxo-1,3,4-thiadiazol-3(2H)-yl] methyl} *O,O*-dimethylphosphorodithioate (9CI). CAS No. [950-37-8].

monocrotophos: (E)-dimethyl 1-methyl-3-(methylamino)-3-oxo-1-propenyl phosphate (9CI). CAS No. [6923-22-4].

paraquat: 1,1'-dimethyl-4,4'-bipyridinium; 1,1'-dimethyl-4,4'-bipyridyldiylium ion (8&9CI). CAS No. [4685-14-7].

parathion: *O,O*-diethyl *O*-(4-nitrophenyl) phosphorothioate (9CI). CAS No. [56-38-2].

parathion, methyl: *O,O*-dimethyl *O*-(4-nitrophenyl) phosphorothioate (9CI). CAS No. [298-00-0].

permethrin: (3-phenoxyphenyl)methyl 3-(2,2-dichloroethenyl)-2,2-dimethylcyclopropanecarboxylate (9CI). CAS No. [52645-53-1].

phosmet: S-[(1,3-dihydro-1,3-dioxo-2H-isoindol-2-yl)methyl]*O,O*-dimethyl phosphorodithioate (9CI). CAS No. [732-11-6].

phosphamidon: 2-chloro-3-(diethylamino)-1-methyl-3-oxo-1-propenyl dimethyl phosphate (9CI). CAS No. [13171-21-6].

pirimiphos, methyl: *O*-[2-(diethylamino)-6-methyl-4-pyrimidinyl] *O,O*-dimethyl phosphorothioate (8&9CI). CAS NO. [29232-93-7].

p,p'-DDT: 1,1'-(2,2,2-trichloroethylidene)*bis*[4-chlorobenzene] (9CI). CAS No. [50-29-3].

simazine: 6-chloro-*N,N'*-diethyl-1,3,5-triazine-2,4-diamine (9CI). CAS No. [122-34-9].

trichlorfon: dimethyl (2,2,2-trichloro-1-hydroxyethyl)phosphonate (8&9CI). CAS No. [52-68-6].

Potential Fate of Chemical Warfare Agents on Kuwait Soil

Hosny Khordagui*

Contents

I. Introduction

The Iran-Iraq war, followed by the Iraqi aggression against Kuwait, and the unverified use of certain chemical warfare agents (CWAs) in the Arabian Gulf region (New Scientist 1985) triggered the interest of environmental scientists on the probable fate of these chemical agents within the unique arid environment of the Arabian Gulf region.

The fate of CWAs in any environment is a very complex process involving physical, chemical, and sometimes biological or biochemical reactions. Once dispensed, the poison will either disintegrate into relatively harmless decay products or will persist in some environmental compartment and represent a continuing threat to living organisms residing in the targeted area (SIPRI 1977). According to Graham-Bryce (1981), the stability, method of dispersion, and physico-chemical properties of the agents in addition to the type of soil, bioavailability, soil organisms, and climate are the major factors determining the behavior and fate of CWAs and/or similar compounds on soil.

Soil provides an environment for chemical changes. Although the soil relative humidity may fall below 90% under extremely dry conditions, soils

*Environmental and Earth Science Division, Kuwait Institute for Scientific Research, P.O. Box 24885, Safat 13109, Kuwait.

© 1995 by Springer-Verlag New York, Inc.
Reviews of Environmental Contamination and Toxicology, Vol. 141.

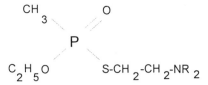

Fig. 1. Structure of the CWA Vx, where R = isopropyl in case of Vx.

in general have a relative humidity of more than 98%, and in such an environment, attack by chemical oxidation and hydrolysis can be expected (Trapp 1985). Of even greater significance, the soil provides an extremely heterogeneous microbiological population conditioned for an existence in an energy-deficient environment. According to Morrill et al. (1985), soil microorganisms probably can degrade every organic compound produced in nature; however, many synthetic compounds do not biodegrade because they lack the necessary enzymatic apparatus or genetic material.

Due to the scarcity of information on the characteristics of the CWAs of interest, their potential methods of application, and the unique, complex nature of the hot and arid soil environment of Kuwait, it would be extremely difficult to quantify any specific interaction between the soil and a given CWA. Even in situations where detailed knowledge of most of the environmental settings and contributory processes are understood, projections concerning the fate and redistribution of nerve CWAs are descriptive and should be treated prudently.

An understanding of the factors affecting such complex processes can lead one to project the minimum duration and conditions needed for detoxification or self-cleaning of the area that has been attacked.

II. Chemical Warfare Agents

Nerve CWAs are considered among the most important lethal agents currently available in the military arsenals of countries in the area and ready for potential application in the Arabian Gulf region (Norman 1989; Zanders 1994). These agents are chemically related to organophosphorus pesticides, but they are much more toxic. At present, the three most active and prevailing nerve CWAs, known as G agents, are (1) Tabun (GA), ethyl phosphorodimethylamidocyanidate; (2) Sarin (GB), isopropyl methylphosphonofluoridate; and (3) Soman (GD), pinacolyl methylphosphonofluoridate.

A prominent class of nerve agents discovered after World War II is the V agents. Generally, these are colorless and odorless liquids, which do not evaporate rapidly at normal temperatures. They are characterized by the S-2-N N-dialkylaminoethyl side chain and are alkylphosphonothiolates with the structure given in Fig. 1. The most important of the V agents is Vx, which is estimated to be three times as potent as the nerve agent Sarin (GB). In the present work, emphasis was given to Vx due to its extremely high potency as compared with other nerve agents.

III. Environmental Fate Processes

The most important environmental fate processes to be considered in the present work are as follows:

A. Hydrolysis

In soil, the reaction illustrated in Fig. 2 represents the decomposition products of a regular hydrolysis reaction of GB. According to Larson (1957), the hydrolysis process proceeds as a one-stage reaction and involves nucleophilic attack of the hydroxyl ion on the phosphorus atom. The attraction between them is controlled by the inductive and mesomeric effects of the substitutes, which also influence the effect of the P–F bond. The yield of hydrolysis in both acidic and basic solutions is two molecules of acids.

The hydrolysis of G agents is principally a function of pH and temperature (Lindesten et al. 1975). At medium pH range, GB has maximum stability toward hydrolysis. According to Epstein (1974), at pH = 7, the primary reaction is very slow, but in unbuffered solution, the acidification due to the production of HF increases the rate of reaction after a phase of stabilization.

The nerve agent GD decomposes in soil via hydrolysis at a slower rate than GB. However, the hydrolysis is comparable with regard to pH and temperature dependence. GD is most stable in the medium pH range from 4 to 6 and is self-buffering in that range. Below pH 4, the hydrolysis is self-catalyzed due to the generation of hydrogen ions and phosphate ions.

In the hydrolysis of the nerve agent GA, not only is there a splitting off of a cyanogen group, but there is at the same time an attack on the P–N bond. This is the only explanation for the great decrease in toxicity of aqueous solutions of GA. The splitting off of the CN group is especially affected by the hydroxyl ions, whereas the formation of the amine is to be

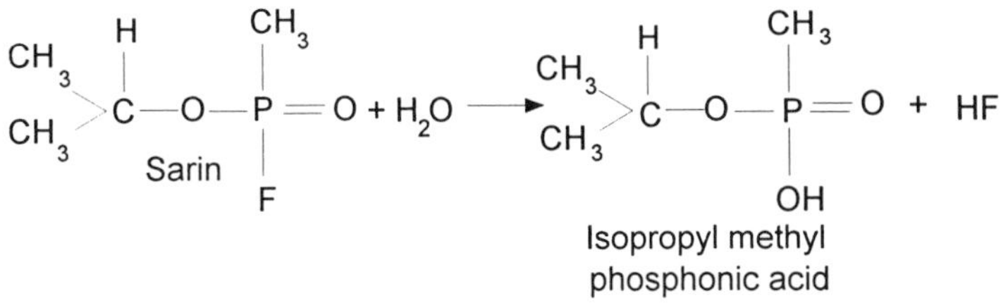

Fig. 2. Hydrolysis of Sarin in soil (after Franke 1977).

attributed to the influence of the hydrogen ions. According to Franke (1977), the hydrolysis of GA takes place as illustrated in Fig. 3.

For the phosphonic esters, such as in the case of Vx, a more general observation with respect to their reactivity is that the introduction of a directly bonded alkyl group onto the central P atom decreases the π electron acceptor capacity of the ester bond. This ultimately leads to an increase in alkaline but a decrease in acidic hydrolysis rate. According to Trapp (1985), the main chemical reactions that phosphonic esters may undergo or take part in are hydrolysis, oxidation, reduction, isomerization, and alkylation. The mechanism of Vx degradation in soil as proposed by Verwej and Boter (1976) is given in Fig. 4. The applied concentration of the Vx agent was 200 mg/kg humic sand, humic loam, and clayey peat soil. The degradation of Vx was fast, as indicated in the proposed mechanism. The concentration of the original compound dropped to 22% for humic sand and to 2% for the other two soils in only 1 d.

The rate of hydrolysis of organic chemicals generally increases with the temperature. The relationship between the hydrolytic constant (k) and temperature is frequently expressed by the Arhenius equation (Yaron 1978) as follows:

$$k = se^{-Ha/RT},$$

where s = frequency factor; Ha = heat of activation (kcal/mole); R = gas constant (1.987 cal/deg mole); T = temperature (K).

In theory, the temperature dependence of k is more complex than the equation would suggest because s and Ha are themselves temperature-dependent. For temperatures ranging from 0 to 50 °C, the expected impact on the hydrolytic rate constant was projected by Harris (1982) to be as

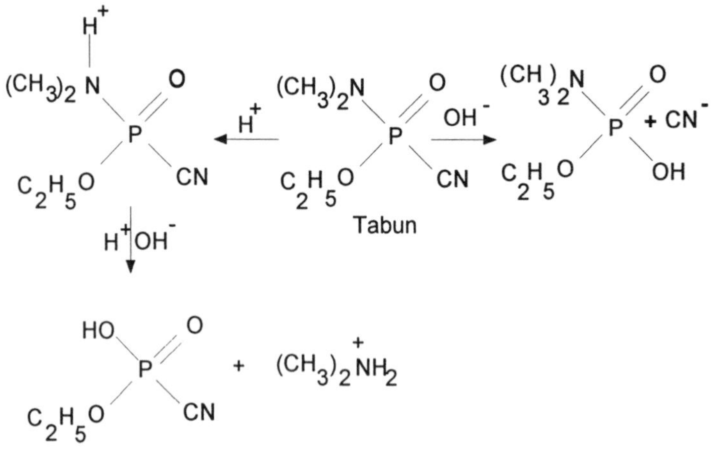

Fig. 3. Mechanism of Tabun hydrolysis (after Franke 1977).

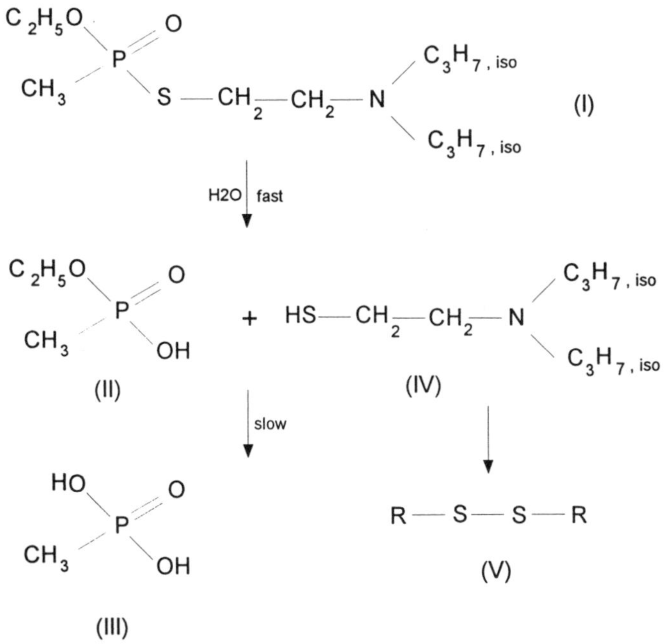

Fig. 4. Mechanisms of Vx degradation in soil as proposed by Verwej and Bottert (1976).

follows: A 1 °C change in temperature will cause a 10% change in k; A 10 °C change in temperature will cause a factor of 2.5 change in k; and A 25 °C change in temperature will cause a factor of 10 change in k.

Because of the high sensitivity of k to changes in soil temperature, a diurnal or seasonal variation of 25 °C would cause a ten-fold (1000%) increase in the hydrolysis rate of the organic compound.

B. Evaporation

Evaporation is considered to be the most important mechanism for the loss of nerve CWA from arid topsoil. The process clearly depends on the chemical properties of the agents, its vapor pressure, and the degree of contamination (Chinn 1981). Some of the processes taking place within the soil subsurface are likely to affect the concentration of the agent at the surface and therefore its rate of evaporation. These processes, such as leaching, diffusion, and adsorption to soil particles, are in turn influenced by the prevailing meteorological conditions, such as temperature, precipitation, wind speed and relative humidity. Soil characteristics, such as clay, organic and moisture contents, porosity, density, and sorptive capacity, all can play an important role in determining the rate of evaporation of the applied CWAs.

The persistence of nerve CWAs on soil under different weather conditions was listed by Franke (1977) as given in Table 1.

C. Adsorption

The amount of nerve CWAs to be adsorbed on a certain soil is determined by the properties of the agent, soil composition, and prevailing environmental and meteorological conditions. Due to the soil system's complexity, the Langmuir adsorption isotherm cannot be used to describe the process. In real time, soil system uptake would cease to be proportional to the concentration of the CWAs because of crowding on the surface of the soil particles.

The major soil characteristics affecting the adsorption phenomena are: (1) percent clay; (2) total soil surface area; (3) pH; (4) cation exchange capacity; (5) percent organic matter; (6) clay mineral type; (7) salt content; (8) water content; (9) percent pore space; and (10) soil temperature. According to Trapp (1985), the factors that are of significance in soil adsorption are the total charge of the soil and its available surface area. The charge referred to is the permanent negative charge of the organic matter/minerals that is balanced by cations. For most of the CWAs, it is the organic matter of the soil that has supreme significance. For adsorption, by far the most important component of the organic matter is the soil content of humic substances.

If the CWA is sorbed while the soil is wet, desorption occurs quickly; however, if the soil is dry for 2 or 3 d after uptake, desorption becomes much more difficult. The influence of pH value of the contaminated soil might change the solubility of the CWAs leading to the formation of charged ions such as oxonium, sulphonium, ammonium, and similar ions. According to Trapp (1985), under certain conditions, most of the CWAs are, capable of forming such charged species. Amino groups as in Vx are important, for, upon protonation, they may become involved in cation exchange and possibly participate in hydrogen bonding. The low solubility of Vx might be used as an indicator for higher adsorption of the compound on soil. Soil temperature affects the adsorption rate in that an increase in

Table 1. Persistence of nerve CWAs under different weather conditions.

Nerve CWA	Sunshine, slight wind at 15 °C	Rainfall, foul, at 10 °C	Sunshine, calm on snow at -10 °C
Sarin	0.5–4 hr	0.25–1 hr	1–2 d
Soman	2.5–5 d	3–36 hr	1–6 wk
Vx	3–21 d	1–12 hr	1–16 wk

temperature generally leads to a decrease in adsorption and/or to an increase in desorption of the CWAs.

An additional and significant phenomenon related to adsorption of nerve CWAs is the potential formation of bond residues. Bond residues are those amounts of CWAs which, after a certain time lapse, cannot be mobilized again either by polar (water) or nonpolar solvents under normal conditions. This effect was noticed for several pesticides, including organophosphorus ones. According to Khan (1980), the structural characteristics of the bounded chemicals may survive and may then be released under certain circumstances. This phenomenon might lead to the protection of the CWAs from chemical and microbial attacks and the long-term persistence in soil toxicity.

D. Photodegradation

Photodegradation applies only to the small fraction of deposited CWA onto the uppermost soil surface, having no relevance for the agent within the soil. Although not important, photodegradation can ultimately affect the fate of CWAs in the environment to a limited extent. Harris (1982) has estimated that photophysical deactivation to the ground state of the CWA molecules, with no net chemical degradation, generally accounts for more than 95% of the light energy absorbed. Thus, it can be concluded that the potential for photolysis of CWAs after dissemination would be insignificant.

E. Biodegradation

Almost all of the reactions involved in microbial degradation can be classified as reduction, oxidation, hydrolysis, or conjugation (Morrill 1985). Most of the research in this field was found to be descriptive, focusing on identification of the organisms responsible for the degradation of specific substances (Morrill 1985; Alexander 1977; Scov 1982). Soil consists of discrete compartments, some being more stable as a microbial habitat. The majority of the microbial population is located in the top layer of soil. As would be expected, microbial density is affected by the organic matter content of soils, which could range from < 1% in mineral soils to > 90% in rich organic soils (Alexander 1977). On the average, the total number of microorganisms in the topsoil is 12×10^6 organisms/g soil, representing some 80% of the total number of organisms counted from various soil depths. According to Scov (1982), the variables that can influence the rate of biodegradation are substrate, organism, and environment related.

Nerve CWAs are orders of magnitude more toxic to mammals than other chemicals that have been tested for microbial degradation (Helling et al. 1971). On the other hand, however, the mechanism of their toxic action (acetylcholinesterase inhibition) is very specific and appears to have no relevance to microbial survival. For compounds closely related both chemically and toxicologically to the nerve gases, it has been shown that soil-microorganisms degradation is a relevant pathway. *Pseudomonas melophthora* and *Pseudomonas testosteroni* were found capable of degrading chemicals containing a carbon–phosphorus bond (alkylphosphonic acid derivatives), including such chemicals as *O*-isopropyl hydrogen methylphosphonate and *O*-pinacolyl hydrogen methylphosphonate, which are the primary metabolites of GB and GD. Metabolites of G-CWAs, namely isopropyl and pinacolyl methylphosphonates, were used as phosphorus sources by *Pseudomonas putida* (Cook et al. 1978)

Generally, few microorganisms found in soil can indeed break down the C–P bond, which is essential for the metabolism of nerve CWAs. The biodegradation process is assumed to be promoted by warm soil temperatures, adequate moisture and the presence of organic matter.

IV. Effects of Environmental Conditions in Kuwait

The unique arid soil characteristics of Kuwait and its surroundings should play an important role in defining the fate of the deposited nerve CWAs. The general topography of the state of Kuwait is flat to gently undulating arid desert plain of low relief. The land rises gradually from the shores of the Arabian Gulf to the extreme southwest of the country with an altitude of approximately 300 m. The desert landscape is frequently broken by elevations, valleys, depressions, sand dunes, and marshes. The climate is characterized by its hot dry summer and cool-to-mild rainy winter. The vegetation is poor open scrub or undershrubs, perennial herbs, and ephemeras primarily controlled by rainfall (EPC 1992).

The geological formations consist of recent marine deposits, outcrops of Miocene rocks, and feldpathic sands and shelly limestones. In defining the distribution of vegetation in Kuwait, annual grasses are not as important as perennial plants because of the very low moisture content of the surface layers of the soils. Ergun (1974) has classified Kuwaits soil as follows:

(1) Desert soils – These are undulating plains formed under arid dry climatic conditions. They are characterized by sand texture and very low organic matter with cemented calcareous subsoil.
(2) Desert intergrade soils – They have been developed under the same conditions as desert soils. They differ from the desert soils in having a coarser texture, better drainage, and the lack of a pan layer.
(3) Lithosols – These are undeveloped, very shallow soils, which have been found on the scrapment, overlaying sedimentary gypsiferous sandstone material.
(4) Alluvial soils – They are found along the Arabian Gulf shoreline. These soils are of marine origin, young, saline, gypsiferous, and lacking definite profile development, and they are wet most of the year due to the high water table.

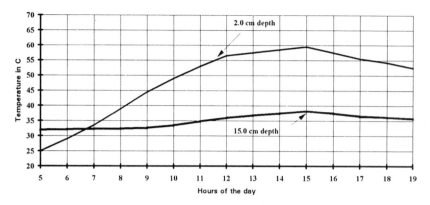

Fig. 5. Daily average summer soil temperature in Kuwait in °C.

A. Temperature and Humidity

In Kuwait, July and August are the hottest months of the year. During these months, the average maximum and minimum temperatures are approximately 45 ° and 29 °C, respectively, whereas the absolute maximum and minimum recorded temperatures are 51° and − 4 °C, respectively. The average temperature in winter varies between 19.4 °C in March and 12.7 °C in January (Al-Ajmi and Safar 1987). Higher ambient temperature raises the soil temperature in Kuwait to record levels. Figures 5 and 6 illustrate the average daily and monthly variations in soil temperature at two different depths in summer and around the year, respectively. Because of the poor thermal conductivity of sandy soil in Kuwait, the energy received by it from the sun is concentrated mainly in the thin top layer, and because of its small heat capacity, there is a large fluctuation in the temperature of the surface layer. Sandy soil has the greatest temperature range in the top layer of any

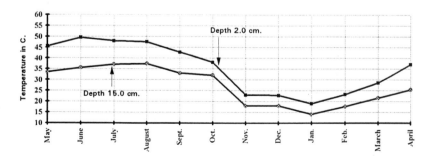

Fig. 6. Average monthly variation of soil temperature in Kuwait in °C.

type of soil, followed by loamy and then clay soils. In addition, the heat capacity of soil varies greatly according to its moisture content.

It is predicted that at these high temperatures, higher rates of evaporation will take place. The evaporation of CWAs from Kuwait soil surface will be significantly accelerated, particularly during the scorching summer months extending from June to September. Furthermore, higher topsoil temperatures will lead to a decrease in the adsorption capacity and an acceleration in the rate of desorption. At high temperatures, the hydrolysis mechanisms of CWAs will be promoted by several orders of magnitude (Harris 1982). However, in Kuwait, the hydrolytic degradation mechanisms might be hindered by the extremely dry conditions, particularly in the summer seasons. Diurnal variation in topsoil temperature in Kuwait can play a major role in defining the rate of dissipation or degradation of nerve CWAs. During the day, the released CWAs will be subject to a much faster rate of degradation as compared to night, when CWA evaporation and hydrolysis will be at their lowest rates and photodegradation is not existent.

In Kuwait, the summer season is almost dry, especially during June, July, and August, whereas the relative humidity on the average varies between 8% and 38%. During winter, relative humidity is much higher, varying between 19% and 87% (Al-Ajmi and Safar 1987).

Figure 7 provides the variations of the average maximum, minimum, and monthly means of percent relative humidity at Kuwait International Airport (KIP) (Al-Ajmi and Safar 1987). Higher relative humidity in Kuwait, such as in winter months, will hinder the CWA volatilization process significantly. However, this hindrance will be countered by a more efficient biodegradation by soil microorganisms and an accelerated rate of hydrolysis for CWAs deposited on the top layer of soil.

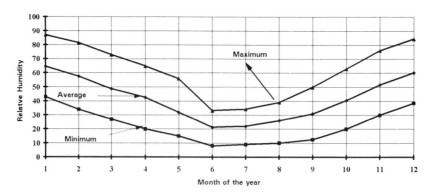

Fig. 7. Monthly maximum, minimum and average percent relative humidity in Kuwait.

B. Rainfall

Precipitation is rare and mainly limited to winter season in Kuwait. The annual rainfall largely varies from year to year. Figure 8 illustrates the monthly minimum, average, and maximum rainfall for the years extending from 1958 to 1987 (Al-Ajmi and Safar 1987).

During the wet seasons, the high water content of the soil will impact the fate of the CWAs via numerous processes known to be of importance in the agent/soil interaction mechanisms. The processes to be affected by the water content in soil are adsorption/desorption, diffusion, hydrolysis, bio-degradation, leaching, surface runoff, and thermal conductivity of soil. From the temporal rainfall distribution and relative humidity patterns in Kuwait, it seems that frequent wetting and drying cycles will be of minimum impact in reducing desorption of CWAs except in the wet and humid winter seasons.

Surface runoff causing lateral movement across the soil surface is very unlikely to happen on Kuwait soil. This is mainly attributed to the limited amount of precipitation per day, the porous nature of the clay-poor soil, and the mostly flat topography of the country.

Biodegradation is also postulated to be promoted during the wet seasons extending from November to March. The increase in biodegradation can be attributed to the expected higher bacterial diversity and better metabolic efficiency. With increasingly higher water content, the diffusion of CWAs in the soil becomes slower, and part of the soil may become completely free of voids. The movement of air or CWA gas into and out of the soil is mainly by diffusion, but there can be mass flow of air near the soil surface at high wind speed. Soil moisture is expected to allow for the hydrolysis

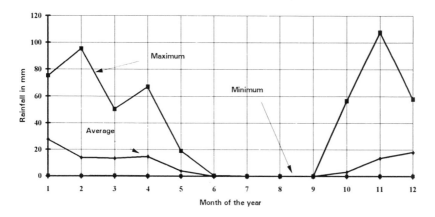

Fig. 8. Monthly maximum, average, and minimum mm rainfall in Kuwait.

reactions of CWAs to proceed. Furthermore, higher thermal conductivity of the wet soil will indirectly lead to a faster rate of hydrolysis.

C. Arid Soil Characteristics

Soil chemical and textural characteristics play a major role in determining the fate of nerve CWAs. According to Ergun (1974), Kuwaits soil can be classified based on its mechanical and chemical composition into a number of series. The topsoil chemical and textural analyses of the different soil series in Kuwait are listed in Table 2.

The hot, mostly sandy, and very dry topsoil covering Kuwait also is characterized by its extremely low organic content, relatively high pH, and low moisture content (Ergun 1974). The temporal distribution of percent soil moisture content around the year is illustrated in Fig. 9.

It is generally postulated that the low organic content, particularly in the form of humus (humic and fulvic acids), in Kuwaits topsoil will limit to a large extent the adsorption and bond residue of nerve CWAs. Low moisture content will promote the diffusion of the CWAs while limiting its mass flow. Areas characterized by dry clay, such as in Bubyan series, when subject to heavy precipitation might lead to a lateral translocation of the CWA across the soil surface. The low percentage of moisture content also might hinder any chemical oxidation and/or hydrolysis from taking place as natural degradation mechanisms for nerve CWAs in soil. On the other

Table 2. Mechanical and chemical analysis for soil series in Kuwait.

Soil series	Depth (cm)	Gravel (%)	Sand (%)	Silt (%)	Clay (%)	Texture class	pH Paste	EC 1000	Cation exchange capacity	CaCO$_3$ equivl. (%)
Kuwait	0–18	7.8	75.9	9.5	6.8	LS	8.2	0.8	3.32	3.32
Ahmadi	0–14	6.2	85.5	4.5	3.8	S	8.4	0.2	4.2	8.04
Um Ar-Rimam	0–5	—	75	13.0	12.0	SL	8.1	1.1	—	16.6
Dibdiba	0–20	—	76	15.0	9.0	SL	8.0	1.0	—	13.9
Sabriya	0–18	—	79	10.0	11.0	SL-LS	7.8	3.5	—	11.8
Batin	0–18	—	82	12.0	6.0	LS	8.1	0.85	—	9.8
Raudhatein	0–9	—	70	13.0	17.0	SL	7.5	2.5	—	16.6
Kra al-Maru	0–30	—	68	19.0	13.0	SL	8.0	2.4	—	16.3
Sadda	0–11	—	68	12.0	20.0	SL-SC	7.8	144	—	10.9
Sulibiya	0–40	—	90	5.0	5.0	S	7.9	1.3	—	11.1
Magwa	0–23	—	89	5.0	6.0	S	8.2	0.68	—	5.2
Juwaisri	0–4	16.4	70.2	8.0	5.4	LS	8.2	0.45	6.3	10.3
Um Nigga	0–25	—	—	—	—	—	8.2	1.5	—	—
Jal Az-Zor	0–5	—	—	—	—	—	7.9	3.3	—	—
Doha	0–6	5.5	70.1	16.3	8.1	SL	7.7	392	d6.4	59.4
Nugaija	0–26	—	73	18	9.0	SL	7.9	66	—	12.6
Ras al-Jilaiya	0–5	3.8	73	12.8	10.4	SL	7.1	176	5.8	9.22
Bubiyan	0–30	—	41.6	22	36.4	CL	7.7	48	—	—

L = loamy; S = sandy; C = clay; — = not determined.

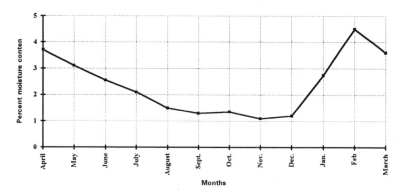

Fig. 9. Average top soil (0–30 cm) percent moisture content in Kuwait.

hand, the relatively high pH of the soil will promote the hydrolysis process in wet seasons.

According to Al-Quadi (1994), the measured average total number of microorganisms in the topsoil of Kuwait is in the range of 10^5 organisms/g soil. This value is two orders of magnitude less than the levels reported by Scov (1982). However, in the very limited reclaimed and fertile cultivated soils in Kuwait, the total number ranges from 10^7 to 10^{14} microorganisms/g soil. Due to the excessively high temperature, low moisture, and organic content, and subsequent low microorganism population in the arid topsoil of Kuwait, it is anticipated that the biodegradation process will have a very limited influence on the elimination of nerve CWAs when released into such an environment.

Summary

The Iran-Iraq war, followed by the Iraqi aggression against Kuwait, and the unverified use of certain chemical warfare agents in the Arabian Gulf region triggered the interest of environmental scientists on the probable fate of these chemical agents within the unique arid environment of the Arabian Gulf region. The basic objective of the present study is to project the effect of local environmental conditions and soil characteristics on the fate processes of nerve chemical warfare agents if deposited on Kuwaiti arid soil.

Open literature indicates that the degradation of nerve CWAs in soil may sometimes be a very effective means for the elimination of these agents. Several of the degradation processes were investigated to some extent. However, several major gaps in the knowledge needed to accurately assess and project the fate of these agents on Kuwait arid soil exist. Among these gaps the following can be listed:

(1) Certain mechanisms are not at all understood and inadequately investigated.

(2) The environmental factors affecting the degradation processes of nerve CWAs in Kuwait are very different and peculiar in terms of soil nature, composition, texture, surface temperature, and moisture and organic content.

(3) Comprehensive understanding of interaction between different simultaneous degradation processes of nerve CWAs in soil as a whole is rather poor today, even in a similar situation where detailed knowledge of most of the contributory processes is available.

(4) A great deal of data acquired by military institutions have not been disclosed and are classified as secret information.

Based on these facts, it is nearly impossible to quantitatively estimate the fate of the CWAs of concern on the arid soil of Kuwait. However, descriptive projections will be very helpful in illustrating the possible fate of these agents as follows:

(1) Surface runoff of nerve CWAs deposited on Kuwait arid soil should be insignificant due to the soil texture, limited precipitation, and topographical features of the country.

(2) Evaporation rate will be accelerated during the windy and exceptionally hot summer season.

(3) Adsorption and bond residue of nerve CWAs in Kuwait soil should be insignificant due to the high topsoil temperature and low organic content.

(4) Hydrolysis of nerve CWAs should be significantly promoted by the characterizing high temperature and pH of the topsoil, particularly when moisture content is adequate for the reaction to proceed.

(5) Biodegradation should be an insignificant process for nerve CWA elimination from arid topsoil in Kuwait. This can be attributed to the low organic and moisture content leading to a decline in the microorganism population.

In general, the only two mechanisms capable of affecting the fate of nerve agents on Kuwait arid soil were evaporation and hydrolysis. Daytime photodegradation and biodegradation were found to have very limited impact.

References

Al-Ajmi DN, Safar MA (1987) An introduction to climatology and climatic geography. Al-Falah Publisher, Kuwait.

Al-Quadi A (1994) Personal communication. Public Authority for Agriculture and Fisheries, Kuwait City, Kuwait.

Alexander M (1977) Introduction to Soil Microbiology, Second ed. John Wiley & Sons, New York.

Anonymous (1985) Gas victims arrive in London. New Scientist 28 (March): 5.

Chinn KSK (1981) A simple method for predicting chemical agent evaporation. Rep DPG-TR-401. U.S. Army Dugway Proving Ground, Dugway, UT.

Cook AM, Daughton CG, Alexander M (1978) Phosphonate utilization by bacteria. J Bacteriology 133(1):85–90.

Environmental Protection Council (EPC) (1992) The National Report, Kuwait.

Epstein J (1974) Properties of Sarin in water, water technology and quality. J Am Water Work Assoc 66:31–37.

Ergun HN (1974) Reconnaissance Soil Survey, Second ed. Ministry of Public Work, Kuwait City, Kuwait.

Franke S (1977) Lehrbuch der militaerchemie. 2 vols. Militaerverlag GDR, Berlin, 512 + 615 pp.

Graham-Bryce IJ (1981) The behavior of pesticides in soil. In: Greenland DJ, Hayes MB (Eds), Chemistry of Soil Processes. John Wiley & Sons, New York.

Harris CR (1982) Factors influencing the effectiveness of soil insecticides. Ann Rev Entomol 17:177–198.

Helling CS, Kearney CS, Alexander M (1971) Behavior of pesticides in soils. Adv Agronomy 23:147–240.

Khan SU (1980) Pesticides in the Soil Environment. Elsevier Scientific, Amsterdam.

Larsson L (1957) The alkaline hydrolysis of isopropoxymethylphosphorylfluoride (Sarin) and some analogues. Acta Chemica Scandinavica, Copenhagen, 11:1131–1142.

Lindesten DC, Schmitt RP (1975) Decontamination of water containing chemical warfare agents. Rep AD-A 012 630, 103 pp. National Technical Information Service, Springfield, VA.

Morrill LG, Reed LW, Chinn KK (1985) Toxic chemicals in the soil environment. Rep 85-8:20:075. Technical Analysis and Information Office, U.S. Army Dugway Proving Ground, Dugway, UT.

Norman C (1989) CIA details chemical weapons spread. Science 243:888.

Scov KM (1982) Rate of biodegradation. In: Layman WJ, Reedl WF, Rosenblatt DH (Eds), Handbook of Chemical Property Estimation Methods. McGraw-Hill Book Co., New York.

SIPRI (1977) Weapons of Mass Destruction and the Environment. Taylor and Francis Ltd, London.

Trapp R (1985) The detoxification and natural degradation of chemical warfare agents, pp. 12–18. Stockholm International Peace Research Institute (SIPRI), Taylor and Francis, London.

Verwej A, Boter HL (1976) Degradation of S-2-diisopropylaminoethyl-o-ethyl-methylphosphonothiolate: Phosphorus combining products, Pesticide Sci 7:355–362.

Yaron B (1977) Some aspects of surface interactions of clays with organophosphorus pesticides. Soil Sci 125:210–216.

Zanders JP (1994) Chemical weapons proliferation in the Gulf region and the strategic balance after Operation Desert Storm. The International Conference on the Effects of Iraqi Aggression on the State of Kuwait, April 2–6, Kuwait.

Manuscript received October 3, 1994; accepted October 26, 1994.

Index